Lecture Notes in Physics

Volume 820

For further volumes:
http://www.springer.com/series/5304

The Lecture Notes in Physics

The series Lecture Notes in Physics (LNP), founded in 1969, reports new developments in physics research and teaching—quickly and informally, but with a high quality and the explicit aim to summarize and communicate current knowledge in an accessible way. Books published in this series are conceived as bridging material between advanced graduate textbooks and the forefront of research and to serve three purposes:

- to be a compact and modern up-to-date source of reference on a well-defined topic
- to serve as an accessible introduction to the field to postgraduate students and nonspecialist researchers from related areas
- to be a source of advanced teaching material for specialized seminars, courses and schools

Both monographs and multi-author volumes will be considered for publication. Edited volumes should, however, consist of a very limited number of contributions only. Proceedings will not be considered for LNP.

Volumes published in LNP are disseminated both in print and in electronic formats, the electronic archive being available at springerlink.com. The series content is indexed, abstracted and referenced by many abstracting and information services, bibliographic networks, subscription agencies, library networks, and consortia.

Proposals should be sent to a member of the Editorial Board, or directly to the managing editor at Springer:

Christian Caron
Springer Heidelberg
Physics Editorial Department I
Tiergartenstrasse 17
69121 Heidelberg/Germany
christian.caron@springer.com

Christian Röthig · Gerd Schön · Matthias Vojta
Editors

CFN Lectures on Functional Nanostructures – Volume 2

Nanoelectronics

Editors
Dr. Christian Röthig
DFG-Centrum für Funktionelle
Nanostrukturen (CFN)
Karlsruhe Institut für Technologie (KIT)
Wolfgang-Gaede-Str. 1a
76131 Karlsruhe
Germany
e-mail: christian.roethig@kit.edu

Prof. Dr. Matthias Vojta
Institut für Theoretische Physik
TU Dresden
01062 Dresden
Germany

Prof. Dr. Gerd Schön
DFG-Centrum für Funktionelle
Nanostrukturen (CFN) and
Institut für Theoretische
Festkörperphysik
Karlsruhe Institut für Technologie (KIT)
Postfach 3640
76021 Karlsruhe
Germany

ISSN 0075-8450
e-ISSN 1616-6361
ISBN 978-3-642-14375-5
e-ISBN 978-3-642-14376-2
DOI: 10.1007/978-3-642-14376-2
Springer Heidelberg Dordrecht London New York

Library of Congress Control Number: 2004110450

Cover design: eStudio Calamar, Berlin/Figueres

Printed on acid-free paper

Springer is part of Springer Science+Business Media (www.springer.com)

Preface

Nanoscience is driven by fascinating scientific challenges, unexpected discoveries, and opportunities to design and control structures and objects on all length scales—from the atomic to the macroscopic. The last decade has witnessed tremendous progress in physics, chemistry, biology, and electrical engineering in creating the necessary tools and techniques, as well as novel theoretical concepts. While some have already termed nanotechnology as one of the key technologies of the twenty-first century, the transition from nanoscience to a nanotechnology has just begun.

The *Center for Functional Nanostructures* (CFN) was founded in July 2001 at the Universität Karlsruhe (TH) and the Forschungszentrum Karlsruhe and is primarily funded by the Deutsche Forschungsgemeinschaft (DFG). Additional funding comes from the State of Baden-Württemberg and from the home institutions, the Universität and the Forschungszentrum now combined to establish the new Karlsruhe Institute of Technology (KIT). The name of the CFN is its program and vision: we want to participate in making the first crucial steps from pure science towards real-world applications, i.e., we aim to realize *functional nanostructures*, concentrating on optical, electronic and biological functions. The CFN covers the following broad research areas:

A Nano-Photonics
B Nano-Electronics
C Molecular Nanostructures
E Nano-Biology
F Nano-Energy.

The CFN is made up of a wide range of research groups from 19 different institutes in Karlsruhe bringing a variety of scientific backgrounds together. The Center thus provides a melting pot where various talents can be combined to address the problems associated with creating functional nanoscale materials. At the same time, the members of the Center are very much aware of the need to develop a common language to facilitate communication amongst the various disciplines.

The miniaturization of electronic devices, which allowed for large-scale integration and ever increasing data-processing rates, has continued into the nanometer range. These developments have stimulated worldwide research activities, and constitute the focus of research area B "Nano-Electronics" of the CFN, where physicists, chemists and electrical engineers in experimental and theoretical groups contribute and collaborate.

Several of the topics of this research area, including single electron transport through quantum dots and nano-devices, magnetic nanostructures, surface physics and tunnelling spectroscopy, noise, quantum dissipation, quantum computing, as well as computational methods, are contained in this Lecture Notes which comprise chapters by CFN members and external lecturers of a CFN summer school.

Outline of this Volume

Ulrich Schollwöck describes recent progress in simulating strongly correlated quantum systems by adaptive time-dependent density matrix renormalization group (DMRG) techniques. DMRG is arguably the most powerful numerical technique for the description of the static properties of strongly correlated one-dimensional bosonic and fermionic systems. An extension of this method to the calculation of the dynamical out-of-equilibrium properties of such quantum systems is presented and, illustrated in the context of spin-charge separation. This new method opens unprecedented possibilities for the calculation of transport properties, of time-dependent Hamiltonians such as in ultra-cold atomic systems, or of decoherence in many-body systems.

Ferdinand Evers and Andreas Arnold describe calculating methods for ab initio calculations of the molecular conductance. Finding the self energy for ab initio transport calculations relevant for molecular electronics can be troublesome. Errors or insufficient approximations made at this step are often the reason why many molecular transport studies become inconclusive. They propose a simple and efficient approximation scheme that follows from interpreting the self energy as an absorbing boundary condition of an effective Schrödinger equation.

Yaroslav M. Blanter provides an overview of recent activities in the field of current noise. After an introduction into the main properties of shot noise in nanostructures, he turns to recent developments, concentrating on issues related to experimental progress, i.e., non-symmetrized cumulants and quantum noise, counting statistics, super-Poissonian noise, current noise, and interferometry.

Alexander Shnirman, Gerd Schön, Ivar Martin, and Yuriy Makhlin describe how Josephson qubits can be used as probes of noise, in particular of $1/f$ noise in the materials used. It builds up on the progress reached recently with quantum mechanical manipulations of solid-state systems. Substantial progress has been achieved with superconducting circuits based on Josephson junctions. They have the advantages that they can be switched quickly and can be integrated into electronic control and measuring circuits. Strong coupling to the external circuits and other parts of the environment brings with it the problem of noise and, thus, decoherence. Therefore, the study of sources of decoherence is necessary. On the other hand, the Josephson qubits have found their first application as sensitive spectrometers of the surrounding noise.

Markus Morgenstern demonstrates in his lecture that scanning tunneling spectroscopy is a versatile tool to measure wave functions within electronic systems. In addition, the spin distribution can be measured by using adequate spin filters for the scanning tip. The principle of the technique is explained and illustrated by several examples using semiconducting and ferromagnetic samples.

Matthias Braun, Jürgen König, and Jan Martinek discuss the possibility to generate, manipulate, and probe single spins in single-level quantum dots coupled to ferromagnetic leads. The spin-polarized currents flowing between dot and leads create a nonequilibrium spin accumulation and polarization of the dot spin. Both the magnitude and the direction of the dot's spin polarization depend on the magnetic properties of leads and their coupling to the dot.

Eduardo R. Mucciolo discusses adiabatic spin pumping with quantum dots. While electronic transport in mesoscopic systems has been studied intensively for the stationary regime, much less is known about phase-coherent non-equilibrium transport when pulses or ac perturbations are used to drive electrons at low temperatures and at small length scales. In this lecture it is shown how open dots can be used to create spin-polarized currents with little or no net charge transfer. The pure spin pump proposed is the analog of a charge battery in conventional electronics and may provide a needed circuit element for spin-based electronics. Other relevant issues such as rectification and decoherence are also discussed.

Alexander W. Holleitner describes spin relaxation in two- and one-dimensional systems. In semiconductors without inversion symmetry, spin–orbit interactions give rise to very effective relaxation mechanisms of the electron spin. Recent theoretical work, based on the dimensionally constrained D'yakonov Perel' mechanism, predicts increasing electron–spin relaxation times for two-dimensional conducting layers with decreasing channel width. The slowing-down of the spin relaxation can be understood as a precursor of the one-dimensional limit.

Detlef Beckmann finally describes electronic transport properties of superconductor–ferromagnet hybrid nanostructures. Recent experimental work is discussed, along with the necessary basic theoretical concepts, and references to more specialized reviews are given for further reading.

Karlsruhe, February 2010

Christian Röthig
Gerd Schön
Matthias Vojta

Contents

Chapter 1
Simulating Strongly Correlated Quantum Systems: Adaptive Time-Dependent Density-Matrix Renormalization Group

Ulrich Schollwöck

1.1 Introduction

The physics of strongly correlated quantum systems continues to pose major challenges in experimental and theoretical physics, ranging from phenomena such as Kondo physics, Luttinger liquid physics, spin chains and ladders through frustrated quantum magnets and high-T_c superconductivity to bulk materials with strong correlations such as transition metal and rare earth compounds. While theory has focused on static, thermodynamic or at most linear-response quantities in the past, recently questions which explicitly involve the out-of-equilibrium time-dependence of such quantum systems have come to the foreground. These questions arise in the context of transport far from equilibrium or of decoherence, particularly as for very small devices mesoscopics and correlation physics merge. However, perhaps the most striking example is provided by the progress in preparing dilute ultracold bosonic and also fermionic alkali gases. Subjected to an optical lattice, these gases are arguably the purest realization of the typical model Hamiltonians of strong correlation physics, such as the Hubbard model [1, 2]. More importantly, the interaction parameters can be tuned experimentally on quantum mechanically relevant time-scales over a huge range, while being known precisely from microscopic calculations. From a theoretician's point of view, this situation is almost ideal, and has stimulated great interest in the development of time-dependent methods.

For linear response, exact diagonalization can provide detailed results for small systems, and quantum Monte Carlo can provide coarse-resolution results after analytic continuation from imaginary time, for systems without the sign problem.

U. Schollwöck (✉)
Department für Physik, LMU München, Theresienstr. 37,
80333 Munich, Germany
e-mail: schollwoeck@lmu.de

M. Vojta et al. (eds.), *CFN Lectures on Functional Nanostructures – Volume 2*,
Lecture Notes in Physics 820, DOI: 10.1007/978-3-642-14376-2_1,

Outside the linear response regime, essentially the only other tool available has been the diagonalization of very small clusters.

In this review, the emphasis is on recent extensions of the density-matrix renormalization group method (DMRG) [3–6] into the real-time domain which make it the currently most powerful method for such problems. Following up on early attempts to extend DMRG to real-time, input from quantum information theory has led to the formulation of two DMRG algorithms for real-time evolutions. We set out with an explanation of the DMRG method, and move on to its extension to the time-domain [7–10]. The range and power of the new method are discussed based on "real-life" applications, where I focus on a real-time observation of the quintessential 1D phenomenon, spin-charge separation.

1.2 Density-Matrix Group Renormalization

Let us start from the fundamental observation that the key to the simulation of quantum systems is the mandatory and desirable compression of information: while the diverging number of degrees of freedom makes compression necessary on finite computing devices, compression also leads to the emergence of macroscopically meaningful concepts such as temperature and pressure which do not have microscopic counterparts. Yet, this divergence of degrees of freedom is of a different type for classical and quantum many-body systems: considering N spins on a lattice, the number of degrees of freedom diverges as $2N$ (2 angles per spin) for a classical vector spin, which is polynomial in N, whereas it diverges as 2^N, i.e., exponentially, for quantum spin-1/2. This is of course due to the presence of superpositions in quantum physics, making simulation exponentially harder.

1.2.1 Decimation of State Spaces

As quantum computers and simulators are not likely to be available in the immediate future, we are stuck to classical computers for the simulation of quantum systems. Essentially three strategies of dealing with the exponentially large state space are being followed: exact diagonalization deliberately limits itself to the exact study of small quantum systems, being limited to less than 40 spins or 20 electrons. Stochastic techniques try to sample the state space efficiently; this is the realm of the quantum Monte Carlo methods, which, however, run into serious trouble for frustrated spins or fermionic systems due to the negative-sign problem. The last group of methods proceeds via a systematic choice of a subspace which is hoped to contain the physically most relevant states. This implies a physically driven process of discarding states, which I will refer to as decimation. All variational and renormalization group techniques are in this group, and the essential question is of course to identify the best decimation strategy which will depend both on the system and the question asked.

Fig. 1.1 One-dimensional arrangement of energy levels along the energy axis

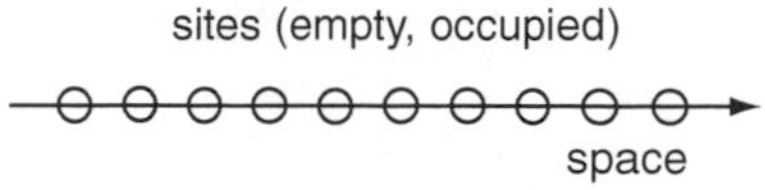

Fig. 1.2 Local lattice degrees of freedom for a one-dimensional system

Let me focus on one particular, rather general decimation setup, "one-dimensional" decimation. Let us assume that the degrees of freedom can be arranged on a 1D axis, e.g., an energy axis, on which we place all levels, either empty or occupied, or a real-space axis, on which we place lattice sites (Figs. 1.1, 1.2).

Imagine we grow the system iteratively towards the thermodynamic limit, adding site by site (whatever the physical interpretation of such a site will be) Fig. 1.3. The original system, which I refer to as a block, is assumed to be effectively described within a state space $\{|\alpha\rangle\}$ of dimension M, the new site within a state space $\{|\sigma\rangle\}$ of dimension N. Obviously, the state space $\{|\beta\rangle\}$ of the new block will have dimension MN, and for prevention of exponential growth it will be decimated down to dimension M. Whatever the physical decimation prescription will be, the states of the new block will be a linear combination of the old states,

$$|\beta\rangle = \sum_{\alpha} \sum_{\sigma} \langle \alpha\sigma|\beta\rangle |\alpha\rangle |\sigma\rangle. \tag{1.1}$$

For later purposes, I will rewrite this expression as

$$|\beta\rangle = \sum_{\alpha} \sum_{\sigma} A_{\alpha\beta}[\sigma] |\alpha\rangle |\sigma\rangle, \tag{1.2}$$

where N matrices A of dimension $M \times M$ have been introduced, one for each $|\alpha\rangle$, such that the matrix entries encode the expansion coefficients: $A_{\alpha\beta}[\sigma] = \langle \alpha\sigma|\beta\rangle$. From the orthonormality of the $|\beta\rangle$ it follows that the A-matrices obey

$$\sum_{\sigma} A^{\dagger}[\sigma] A[\sigma] = 1. \tag{1.3}$$

Finding these matrix entries (i.e., decimating) in the optimal way is of course desirable; but before we try to formulate this as a well-posed problem, it is convenient to observe the following: the states of the block of length, say, $\ell - 1$ serving as "input" to produce as "output" the states of the block of length ℓ via the

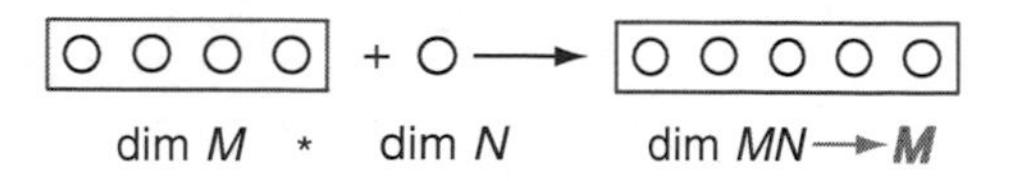

Fig. 1.3 Growing a system: a local state space of dimension N is added to a block state space of dimension M

matrices $A^{\ell}[\sigma_\ell]$, are of course themselves the "output" for the construction of the block of length $\ell - 1$ from that of $\ell - 2$, and so forth. We can therefore recursively write $|\beta\rangle$ for a block of length ℓ as

$$|\beta\rangle = \sum_{\sigma_1\ldots\sigma_\ell} [A^1[\sigma_1]A^2[\sigma_2]\ldots A^{\ell}[\sigma_\ell]]_\beta|\sigma_1\ldots\sigma_\ell\rangle. \tag{1.4}$$

Obviously, for very short blocks, there are less than M states in a block, and the dimensions of the first A-matrices will be $1 \times N$, $N \times N^2$, ... till M is exceeded.

It is now convenient to consider also what happens if we start growing a chain from its right end (site L). Obviously, similar expressions emerge: for a block of length ℓ the states are given as

$$|\gamma\rangle = \sum_{\sigma_{L-\ell+1}\ldots\sigma_L} [\tilde{A}^{L-\ell+1}[\sigma_{L-\ell+1}]\tilde{A}^{L-\ell+2}[\sigma_{L-\ell+2}]\ldots\tilde{A}^{L}[\sigma_L]]_\gamma|\sigma_{L-\ell+1}\ldots\sigma_L\rangle. \tag{1.5}$$

Again, at the chain end the $\tilde{A}$ matrices (defined in analogy to the A) have reduced dimensions instead of $M \times M$; there is also a slightly modified orthonormality constraint,

$$\sum_\sigma \tilde{A}[\sigma]\tilde{A}^\dagger[\sigma] = 1. \tag{1.6}$$

We may now choose a "left" and a "right" block to join them to the generic description of a quantum state of the system of length L, matching them directly,

$$|\psi\rangle = \sum_{\sigma_1\ldots\sigma_L} A^1[\sigma_1]\ldots A^{\ell}[\sigma_\ell]\Psi\tilde{A}^{\ell+1}[\sigma_{\ell+1}]\ldots\tilde{A}^{L}[\sigma_L]|\sigma_1\ldots\sigma_L\rangle, \tag{1.7}$$

where Ψ is a $M \times M$ matrix, or with one site explicitly in between,

$$|\psi\rangle = \sum_{\sigma_1\ldots\sigma_L} A^1[\sigma_1]\ldots A^{\ell}[\sigma_\ell]\Psi[\sigma_{\ell+1}]\,\tilde{A}^{\ell+2}[\sigma_{\ell+2}]\ldots\tilde{A}^{L}[\sigma_L]|\sigma_1\ldots\sigma_L\rangle, \tag{1.8}$$

where there is a Ψ-matrix for each $|\sigma_{\ell+1}\rangle$, or with two sites explicitly in between,

$$|\psi\rangle = \sum_{\sigma_1\ldots\sigma_L} A^1[\sigma_1]\ldots A^{\ell}[\sigma_\ell]\Psi[\sigma_{\ell+1}\sigma_{\ell+2}]\,\tilde{A}^{\ell+3}[\sigma_{\ell+3}]\ldots\tilde{A}^{L}[\sigma_L]|\sigma_1\ldots\sigma_L\rangle. \tag{1.9}$$

As it turns out, (1.8) and (1.9) are the useful ones—both of them highlighting different aspects of DMRG, leading to what is called single-site and (more conventional) two-site DMRG.

Considering (1.7), obviously the junction can be shifted freely, even fictitiously to the right end of the chain, with all $\tilde{A}$-matrices disappearing. The generic description of a quantum state emerging from a decimation procedure is therefore given by

$$|\psi\rangle = \sum_{\sigma_1\ldots\sigma_L} A^1[\sigma_1]A^2[\sigma_2]\ldots A^{L}[\sigma_L]|\sigma_1\ldots\sigma_L\rangle. \tag{1.10}$$

In fact, there is no need to worry about this product of matrices yielding a scalar factor because of the dimension of the matrices at the right end: manipulations introduced further down (namely the Schmidt decomposition) can be used to show that without any loss the A-matrices shrink to dimensions mirroring those at the left end, e.g., $N^2 \times N$ and $N \times 1$ for the last two. The generic decimation state encoded as in (1.10) is referred to as a *matrix product state*; such states have been studied for a long time [11–13].

Again, the A-matrices encode the physical decimation prescription, and we may now specify our question, what is the optimal decimation prescription, for the case of finding the *ground state* of a Hamiltonian. Clearly, the answer is to find the prescription yielding those A that minimize $\langle\psi|\hat{H}|\psi\rangle$ assuming normalization. Working out this expression yields a highly non-linear expression for the energy in A, which is numerically close to useless.

Can we find a method that turns this into a linear problem in A, which would be much easier to implement in a stable way on a computer? In fact, we can, and this has already been done in the past in the form of the density-matrix renormalization group [3–5]. Its link to matrix product states has been pointed out by various authors [14–17]. As there are many reviews of DMRG following a traditional way of understanding (e.g., [5]), going to more depth than possible here, I want to present a different viewpoint of the method as an answer to the question above.

A way of turning the problem linear would consist, roughly speaking, in providing some starting set of matrices in a warm-up procedure, preferably close to the true solution, and then to proceed iteratively: keeping all elements (i.e., matrices) in $|\psi\rangle$ fixed, with one exception, $\langle\psi|\hat{H}|\psi\rangle$ turns quadratic in the free matrix, and extremizing this expression with respect to the free matrix is a linear procedure. Varying one element after another repeatedly, one may hope to get a very good, even optimal, approximation of the ground state within the state class expressable as above.

In standard DMRG language, the first warm-up part of the procedure would correspond to the so-called infinite-system algorithm, the second to the finite-system algorithm, which is the true cornerstone of the method. As the infinite-system algorithm is described very well in many references, and one may even imagine starting from random matrices, I will focus on the finite-system algorithm.

1.2.2 Finite-System DMRG Procedures

1.2.2.1 Two-site DMRG

Let us consider we have our quantum state in the representation of (1.9). In that case, the left block of length ℓ formed from A-matrices would be described by a M-dimensional Hilbert space with states $\{|m_\ell^S\rangle\}$ defined as in (1.4); similarly the right block of length $L - \ell - 2$ by states $\{|m_{L-\ell-2}^E\rangle\}$. In both cases, we assume that we know all necessary operators on the blocks in these effective bases. We can therefore construct the Hamiltonian acting on

$$\begin{aligned} |\psi\rangle &= \sum_{m^S=1}^{M}\sum_{\sigma^S=1}^{N}\sum_{\sigma^E=1}^{N}\sum_{m^E=1}^{M} \Psi_{m^S m^E}[\sigma^S\sigma^E]|m^S\sigma^S\rangle|m^E\sigma^E\rangle \\ &\equiv \sum_{i}^{N^S}\sum_{j}^{N^E} \Psi_{ij}|i\rangle|j\rangle; \quad \langle\psi|\psi\rangle = 1, \end{aligned} \tag{1.11}$$

where $\Psi_{m^S m^E}[\sigma^S\sigma^E] = \langle m^S\sigma^S;\sigma^E m^E|\psi\rangle$. $\{|m^S\sigma^S\rangle\} \equiv \{|i\rangle\}$ and $\{|m^E\sigma^E\rangle\} \equiv \{|j\rangle\}$. The state spaces $\{|i\rangle\}$ and $\{|j\rangle\}$have dimensions $N^S = MN$ and $N^E = MN$. Using some large sparse matrix solver one may minimize the energy of $|\psi\rangle$ with respect to Ψ_{ij}.

In order to make progress, we now have to shift the position of the two active sites, to improve our wave function everywhere. If we shift it by one site to the right, the shrinking right block may easily be constructed from $\tilde{A}$-matrices; for the growing left block, we have to provide A-matrices for the left of the two sites (states $|\sigma^s\rangle$). If one has a suitable truncation of the basis $\{|i\rangle\}$ down to M states, their expansion in $|m^S\rangle$ and $|\sigma^s\rangle$ will just define the desired A-matrices.

To find that truncation, a useful concept is that of a *Schmidt decomposition:* Consider a quantum state $|\psi\rangle = \sum_{ij}\psi_{ij}|i\rangle \otimes |j\rangle$ as introduced before, with N^S states $|i\rangle$ and N^E states $|j\rangle$. Assuming without loss of generality $N^S \geq N^E$, we form the $(N^S \times N^E)$-dimensional matrix A with $A_{ij} = \psi_{ij}$. Singular value decomposition guarantees $A = UDV^T$, where U is $(N^S \times N^E)$-dimensional with orthonormal columns, D is a $(N^E \times N^E)$-dimensional diagonal matrix with non-negative entries $D_{\alpha\alpha} = \sqrt{w_\alpha}$, and V^T is a $(N^E \times N^E)$-dimensional unitary matrix; $|\psi\rangle$ can be written as

$$\begin{aligned} |\psi\rangle &= \sum_{i=1}^{N^S}\sum_{\alpha=1}^{N^E}\sum_{j=1}^{N^E} U_{i\alpha}\sqrt{w_\alpha}V^T_{\alpha j}|i\rangle|j\rangle \\ &= \sum_{\alpha=1}^{N^E}\sqrt{w_\alpha}\left(\sum_{i=1}^{N^S}U_{i\alpha}|i\rangle\right)\left(\sum_{j=1}^{N^E}V_{j\alpha}|j\rangle\right). \end{aligned} \tag{1.12}$$

The orthonormality properties of U and V^T ensure that $|w^S_\alpha\rangle = \sum_i U_{i\alpha}|i\rangle$ and $|w^E_\alpha\rangle = \sum_j V_{j\alpha}|j\rangle$ form orthonormal bases of system and environment respectively, in which the Schmidt decomposition

$$|\psi\rangle = \sum_{\alpha=1}^{N_{\text{Schmidt}}} \sqrt{w_\alpha}|w^S_\alpha\rangle|w^E_\alpha\rangle \tag{1.13}$$

holds. $N^S N^E$ coefficients ψ_{ij} are reduced to $N_{\text{Schmidt}} \leq N^E$ non-zero coefficients $\sqrt{w_\alpha}$,$w_1 \geq w_2 \geq w_3 \geq \ldots$. Relaxing the assumption $N^S \geq N^E$, one has

$$N_{\text{Schmidt}} \leq \min(N^S, N^E). \tag{1.14}$$

Upon tracing out environment or system the reduced density matrices for system and environment are found to be

$$\hat{\rho}_S = \sum_{\alpha}^{N_{\text{Schmidt}}} w_\alpha |w_\alpha^S\rangle\langle w_\alpha^S|; \quad \hat{\rho}_E = \sum_{\alpha}^{N_{\text{Schmidt}}} w_\alpha |w_\alpha^E\rangle\langle w_\alpha^E|. \tag{1.15}$$

Analyzing reduced density matrices or the Schmidt decomposition therefore yield exactly the same information. This fact was understood from the very beginning of DMRG, although DMRG people were not aware of the term "Schmidt decomposition". In fact, the singular value decomposition representation of the wavefunction was understood before the density matrix representation. How can we put this information to good use? Interestingly enough, it allows us to define several physically motivated truncation criteria, that lead to the identical truncation prescription.

1.2.2.2 Optimization of the Wave Function

Quantum mechanical objects are completely described by their wave function. It is thus a reasonable demand for a truncation procedure that the approximative wave function $|\tilde{\psi}\rangle$ where the system space has been truncated to be spanned by only M orthonormal states $|\alpha\rangle = \sum_i u_{\alpha i}|i\rangle$,

$$|\tilde{\psi}\rangle = \sum_{\alpha=1}^{M}\sum_{j=1}^{N^E} a_{\alpha j}|\alpha\rangle|j\rangle, \tag{1.16}$$

minimizes the distance in the quadratic norm

$$\| \, |\psi\rangle - |\tilde{\psi}\rangle \, \| \, . \tag{1.17}$$

Using (1.13), one finds that this distance is minimized if one *retains the* M *eigenstates of* $\hat{\rho}_S$ *with the largest eigenvalues* w_α. This is the key truncation prescription.

One could also ask for maximizing the retained *bipartite entanglement* between system and environment under truncation. As bipartite entanglement is defined as $S = -\sum_\alpha w_\alpha \log_2 w_\alpha$, and that typically one has a large number of relatively small eigenvalues, this again leads to the same truncation prescription, and the method preserves entanglement as well as it can.

Interestingly enough, one can show that the (additional) error introduced by truncation on some generic bounded operator $\hat{A}$ acting on the system, such as the energy per lattice bond, $\|\hat{A}\| = \max_\phi |\langle\phi|\hat{A}|\phi\rangle/\langle\phi|\phi\rangle| \equiv c_A$, is minimized by the above procedure. This error for $\langle\hat{A}\rangle$ is bounded by

$$|\langle\hat{A}\rangle_{\text{approx}} - \langle\hat{A}\rangle| \leq \left(\sum_{\alpha > M}^{MN} w_\alpha\right) c_A \equiv \epsilon_\rho c_A, \tag{1.18}$$

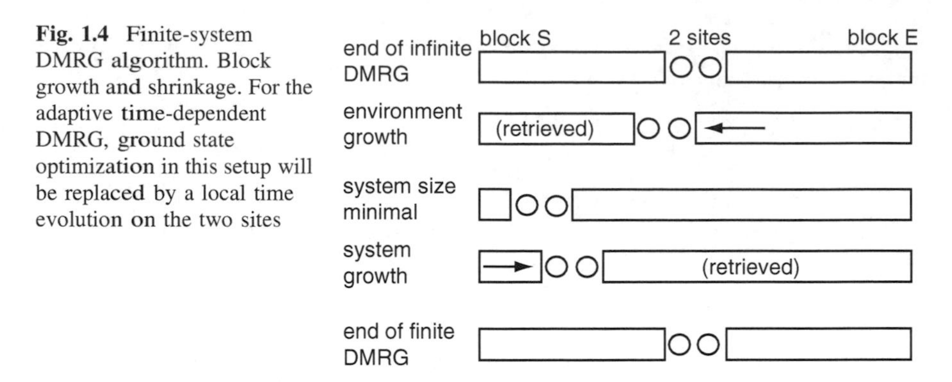

Fig. 1.4 Finite-system DMRG algorithm. Block growth and shrinkage. For the adaptive time-dependent DMRG, ground state optimization in this setup will be replaced by a local time evolution on the two sites

neglecting a small normalization correction. For local quantities, such as energy, magnetization or density, errors are of the order of the *truncated weight*

$$\epsilon_\rho = 1 - \sum_{\alpha=1}^{M} w_\alpha, \tag{1.19}$$

which emerges as the key estimate. Careful extrapolation of results in M (better ϵ_ρ) is therefore highly recommended.

Collecting these ideas, the so-called *finite-system DMRG algorithm* can now be formulated (Fig. 1.4). For a one-dimensional lattice of total length L:

1. Consider a (left) system block S of length ℓ. S lives on a Hilbert space of size M with states $\{|m_\ell^S\rangle\}$; the Hamiltonian $\hat{H}_\ell^S$ and the operators acting on the block are assumed to be known in this basis. Similarly, we have a (right) environment block E of size $L - \ell - 2$.
2. Build the total system (superblock) of length L from S, E and two sites in between. S and the left site live on a Hilbert space of size $N^S = MN$, with a basis of product states $\{|m_\ell^S \sigma^S\rangle\}$. Similarly, environment and the right site are joined. The total Hilbert space is of size $N^S N^E$, and a total Hamiltonian can be constructed.
3. Find by large sparse-matrix diagonalization of the total Hamiltonian the ground state $|\psi\rangle$. This is the most time-consuming part of the algorithm. For better performance, there is a simple, but powerful "prediction algorithm" [18], which cuts down calculation times in finite-system DMRG by more than an order of magnitude. Using the A-matrices, it transforms $|\psi\rangle$ from the last step approximatively into the new basis and uses it as a guess for the next, improved $|\psi\rangle$.
4. Form the reduced density-matrix $\hat{\rho} = \text{Tr}_E|\psi\rangle\langle\psi|$ and determine its eigenbasis $|w_\alpha\rangle$ ordered by descending eigenvalues (weight) w_α. Form a new (reduced) basis for S and the left site by taking the M eigenstates with the largest weights, defining the A-matrices for this site. Proceed likewise for the environment.
5. Form a new system block from S and one added site. As environment block, use the one shorter than E by one site. This effectively shifts the free sites by one to the right.

6. Carry out the reduced basis transformations of needed operators onto the new M^S-state basis. Restart with step (2) with block size $\ell + 1$ until the right end of the chain is reached.
7. Continue the above procedure with system and environment in reversed roles, the free sites being shifted left, until the left end is reached, and so forth, until results are converged. A complete shrinkage and growth sequence for both blocks is called a *sweep*.

DMRG practitioners usually adopt a quite pragmatic approach when applying DMRG to study some physical system. They consider the convergence of DMRG results under tuning the standard DMRG control parameters, system size L, size of the reduced block Hilbert space M, and the number of finite-system sweeps, and judge DMRG results to be reliable or not.

1.2.2.3 Single Site DMRG

In practical applications one observes that even for translationally invariant systems with periodic boundary conditions and repeated applications of finite-system sweeps the position dependency of the matrix-product state (or of observables) does not go away completely as it strictly should, indicating room for further improvement. Also, the method cannot be strictly variational: the ansatz (1.9) generates (after the Schmidt decomposition and before truncation) ansatz matrices of dimension MN at the two local sites due to (1.14), whereas they are of dimension M at all other sites; as the position of these two sites is shifting, this anomaly changes place such that one is not really optimizing within one fixed variational ansatz throughout the algorithm.

Several authors [15, 16] have pointed out and numerically demonstrated that DMRG results can be improved by switching, after convergence is reached, from the S••E scheme for the finite-system algorithm to a S•E scheme as in ansatz (1.8) and to carry out some more sweeps. This is now a truly variational ansatz [15–17].

In the new scheme the ansatz matrix Ψ can be projected down to finding M states for defining the A-matrix without truncation or loss of information or change in energy, because the Schmidt number of non-zero contributions to the density matrix cannot exceed M, which is the dimension of the environment in (1.14) for the single-site setup. Shifting the "active" site then does not change the energy, and the next minimization can only decrease the energy (or keep it constant). This setup is therefore truly variational in the space of the states generated by the matrices A, $\tilde{A}$ and reaches a minimum of energy within that space. It is important to note that in this setup there is therefore no truncation error as diagnostic tool.

There is of course no guarantee to reach the global minimum: in fact, it turns out that an immediate application of the formally superior single site DMRG may fail quite drastically by trapping, but this can be mended by suitable modifications to the state selection scheme, taking the state out of false minima [19].

1.2.3 When and Why Does DMRG Work?

Obviously, the ordered eigenvalue spectrum w_α of the reduced density-matrix $\hat{\rho}$ should decay as quickly as possible to minimize the truncated weight $\epsilon_\rho = 1 - \sum_{\alpha=1}^{M} w_\alpha$ for optimal DMRG performance. The optimal case is of course when the state to be approximated is a one-dimensional $M \times M$ *matrix-product state*, which can be modelled exactly. More generally, in one dimension, density-matrix spectra of gapped one-dimensional quantum systems exhibit roughly system-size independent exponential decay of w_α; at criticality, this decay slows down with increasing system size, leading to DMRG failure for $L \to \infty$. In two dimensions, in reduced density-matrix spectra for states of systems both at and away from criticality, the number of eigenvalues to be retained to keep a fixed truncation error grows exponentially with system size, restricting DMRG to very small system sizes (typically, the two-dimensional lattice is mapped to a one-dimensional snake with long-ranged interactions).

These empirical observations can be understood from examining the growth of bipartite entanglement between system and environment for various dimensions. Consider the entanglement S_L for systems of length L embedded in a thermodynamic limit universe. One finds [20] that $S_L \to \infty$ for $L \to \infty$ at criticality, but saturates as $S_L \to S_L^*$ for $L \approx \xi$ in the non-critical regime. At criticality the entanglement can be linked [20, 21] to the geometric entropy [22] of associated conformal field theories,

$$S_L^{\mathrm{geo}} = \frac{c + \overline{c}}{6} \log_2 L, \tag{1.20}$$

where c ($\overline{c}$) are the central charges. As examples, for the anisotropic XY model $c = \overline{c} = 1/2$ and for the Heisenberg model and isotropic XY model $c = \overline{c} = 1$.

Geometric entropy arguments for $(d+1)$-dimensional field theories use a bipartition of d-dimensional space by a $(d-1)$-dimensional hypersurface, which is shared by system S and environment E. By the Schmidt decomposition, S and E share the same reduced density-matrix spectrum, hence entanglement entropy, which is now argued to reside essentially on the shared hypersurface; see also [23]. Taking the thermodynamic (infrared) limit, entropy scales as the hypersurface area,

$$S_L \propto \left(\frac{L}{\lambda}\right)^{d-1}, \tag{1.21}$$

where λ is some ultraviolet cutoff which in condensed matter physics we may fix at some lattice spacing; critical systems will have logarithmic corrections to (1.21).

S_L is the number of qubits corresponding to the entanglement information. To code this information in DMRG, one needs a system Hilbert space of size $M \geq 2^{S_L}$, we may therefore expect, in perfect agreement with empirical results, that in 1D quantum systems away from criticality, DMRG yields very precise results for the

TD limit for some finite number of states kept, $M \sim 2^{S_L^*}$. At criticality, the number of states that has to be kept, will diverge as

$$M(L) \sim L^k, \tag{1.22}$$

with k inferred from (1.20). This explains the failure of DMRG for critical one-dimensional systems as $L \to \infty$. As k is small, this statement has to be qualified; DMRG still works for rather large finite systems.

In higher-dimensional quantum systems, however, the number of states to be kept will diverge as

$$M(L) \sim 2^{L^{d-1}}, \tag{1.23}$$

rendering the understanding of thermodynamic limit behavior by conventional DMRG quite impossible. In any case, even for higher-dimensional systems, DMRG may be a very useful method as long as system size is kept resolutely finite, such as in nuclear physics or quantum chemistry applications. Recent proposals [24] also show that it is possible to formulate generalized DMRG ansatz states in such a way that entropy shows correct size dependency in two-dimensional systems.

1.3 Time-Dependence in Quantum Systems

Even though the methods described in the previous section provide high-quality linear-response quantities, they fail in truly out-of-equilibrium situations or for time-dependent Hamiltonians; where they work, they are very time-consuming. It has therefore been of high interest to find DMRG approaches dealing with state evolution in real-time.

To see the advantages of such an approach, consider the following. Essentially all physical quantities of interest involving time can be reduced to the calculation of either *equal-time* n-point correlators such as the (1-point) density

$$\langle n_i(t)\rangle = \langle\psi(t)|n_i|\psi(t)\rangle = \langle\psi|e^{\mathrm{i}Ht}n_i e^{-\mathrm{i}Ht}|\psi\rangle \tag{1.24}$$

or *unequal-time* n-point correlators such as the (2-point) real-time Green's function

$$G_{ij}(t) = \langle\psi|c_i^\dagger(t)c_j(0)|\psi\rangle = \langle\psi|e^{+\mathrm{i}Ht}c_i^\dagger e^{-\mathrm{i}Ht}c_j|\psi\rangle. \tag{1.25}$$

This expression can be cast in a form very close to (1.24) by introducing $|\phi = c_j|\psi\rangle$ such that the desired correlator is then simply given as an equal-time matrix element between two time-evolved states,

$$G_{ij}(t) = \langle\psi(t)|c_i^\dagger|\phi(t)\rangle. \tag{1.26}$$

If both $|\phi(t)\rangle$ and $|\psi(t)\rangle$ can be calculated, a very appealing feature of this approach is that $G_{ij}(t)$ can be evaluated in *a single calculation* for all i and t as time proceeds. Frequency-momentum space is then reached by a double Fourier transformation. Obviously, finite system-sizes and edge effects as well as algorithmic constraints will impose physical constraints on the largest times and distances $|i-j|$ or minimal frequency and wave vectors resolutions accessible. Nevertheless, this approach might emerge as a very attractive alternative to the current very time-consuming calculations of $G(k, \omega)$ using the dynamical DMRG [25, 26].

The fundamental difficulty of obtaining the above correlators becomes obvious if we examine the time-evolution of the quantum state $|\psi(t=0)\rangle$ under the action of some (for simplicity) time-independent Hamiltonian $\hat{H}|\psi_n\rangle = E_n|\psi_n\rangle$. If the eigenstates $|\psi_n\rangle$ are known, expanding $|\psi(t=0)\rangle = \sum_n c_n|\psi_n\rangle$ leads to the well-known time evolution

$$|\psi(t)\rangle = \sum_n c_n \exp(-\mathrm{i}E_n t)|\psi_n\rangle, \tag{1.27}$$

where the modulus of the expansion coefficients of $|\psi(t)\rangle$ is time-independent. A sensible Hilbert space truncation is given by a projection onto the large-modulus eigenstates. In strongly correlated systems, however, we usually have no good knowledge of the eigenstates. Instead, one uses some orthonormal basis with unknown eigenbasis expansion, $|k\rangle = \sum_n a_{kn}|\psi_n\rangle$. The time evolution of the state $|\psi(t=0)\rangle = \sum_k \mathrm{d}_k(0)|k\rangle$ then reads

$$|\psi(t)\rangle = \sum_k \left(\sum_n \mathrm{d}_k(0) a_{kn} e^{-\mathrm{i}E_n t} \right) |k\rangle \equiv \sum_k \mathrm{d}_k(t)|k\rangle, \tag{1.28}$$

where the modulus of the expansion coefficients $\mathrm{d}_k(t)$ is *time-dependent*. For a general orthonormal basis, Hilbert space truncation at one fixed time (i.e., $t=0$) will therefore not ensure a reliable approximation of the time evolution. Also, energy *differences* matter in time evolution due to the phase factors $e^{-\mathrm{i}(E_n-E_{n'})t}$ in $|\mathrm{d}_k(t)|^2$. Thus, a good approximation to the low-energy Hamiltonian alone (as provided by DMRG) is of limited use.

1.4 Early Attempts in DMRG

Cazalilla and Marston [27] were the first to exploit DMRG to calculate time-dependent quantum many-body effects. They studied a time-dependent Hamiltonian $\hat{H}(t) \equiv \hat{H}(0) + \hat{V}(t)$, where $\hat{V}(t)$ encodes the time-dependent part of the Hamiltonian. After applying a standard DMRG calculation to the Hamiltonian $\hat{H}(t=0)$, the time-dependent Schrödinger equation was numerically integrated forward in time. The effective Hamiltonian in the reduced Hilbert space was built as $\hat{H}_{\mathrm{eff}}(t) = \hat{H}_{\mathrm{eff}}(0) + \hat{V}_{\mathrm{eff}}(t)$, where $\hat{H}_{\mathrm{eff}}(0)$ was taken as the superblock Hamiltonian approximating $\hat{H}(0) \cdot \hat{V}_{\text{eff}}(t)$ as an approximation to $\hat{V}$ was built using the representations of operators in the superblock bases. The initial condition was obviously to take $|\psi(0)\rangle$ as the ground state obtained by the preliminary DMRG run. This procedure amounts to working within a *static* reduced Hilbert space, namely that optimal at $t = 0$, and projecting all wave functions and operators onto it.

In this approach the hope is that an effective Hamiltonian obtained by targeting the ground state of the $t = 0$ Hamiltonian is capable to catch the states that will be visited by the time-dependent Hamiltonian during time evolution. This approach must however break down after relatively short times as the full Hilbert space is explored, as became quickly obvious.

1.4.1 Dynamic Time-Dependent DMRG

Several attempts have been made to improve on static time-dependent DMRG by enlarging the reduced Hilbert space using information on the time-evolution, such that the time-evolving state has large support on that *dynamic* Hilbert space for longer times. Whatever procedure for enlargement is used, the problem remains that the number of DMRG states M grows with the desired simulation time as they have to encode more and more different physical states. As in DMRG CPU time scales as M^3 (due to the matrix–matrix multiplications involved), this type of approach becomes numerically very expensive.

All enlargement procedures rest on the ability of DMRG to describe—at some numerical expense—small sets of states ("target states") very well instead of just one.

The first approach has been demonstrated by Luo et al. [28]. They use a density matrix that is given by a superposition of states $|\psi(t_i)\rangle$ at various times of the evolution, $\hat{\rho} = \sum_{i=0}^{N_t} \alpha_i |\psi(t_i)\rangle\langle\psi(t_i)|$ with $\sum \alpha_i = 1$ for the determination of the reduced Hilbert space. Of course, these states are not known initially; it was proposed by them to start within the framework of infinite-system DMRG from a small DMRG system and evolve it in time. For a very small system this procedure is exact. For this system size, the state vectors $|\psi(t_i)\rangle$ are used to form the density matrix. This density matrix then determines the reduced Hilbert space for the next larger system, taking into account how time-evolution explores the Hilbert space for the smaller system. One then moves on to the next larger DMRG system where the procedure is repeated. This is of course very time-consuming.

Schmitteckert [29] has computed the transport through a small interacting nanostructure using an Hilbert space enlarging approach, based on the time evolution operator. He implements the matrix exponential $|\psi(t + \Delta t)\rangle = \exp(-i\hat{H}\Delta t)|\psi(t)\rangle$ using the Krylov subspace approximation. For any block-site configuration during sweeping, he evolves the state in time, obtaining $|\psi(t_i)\rangle$ at fixed times t_i. These are targeted in the density matrix, such that upon sweeping forth and back a Hilbert space suitable to describe all of them at good precision should be obtained. Again, this is a very time-consuming approach.

1.5 Adaptive Time-Dependent DMRG

Decisive progress came from an unexpected corner, namely quantum information theory, when Vidal proposed an algorithm for simulating quantum time evolutions of one-dimensional systems efficiently on a classical computer [7, 30]. His algorithm, known as TEBD [time-evolving block decimation] algorithm, is based on matrix product states [12, 13]; as it turned out, it is so closely linked to DMRG concepts, that his ideas could be implemented easily into DMRG, leading to an *adaptive* time-dependent DMRG [8, 9], where the DMRG state space adapts itself in time to the time-evolving quantum state. One immediately profits from all the DMRG development for exploiting good quantum numbers and other speed-ups.

If we consider a nearest-neighbor Hamiltonian, such as the conventional Hubbard Hamiltonian, we can split the infinitesimal global time evolution operator into a product of infinitesimal local time evolution operators [31]:

$$e^{-\mathrm{i}\hat{H}\Delta t} = e^{-\mathrm{i}h_1}e^{-\mathrm{i}h_2\Delta t}e^{-\mathrm{i}h_3\Delta t}\dots e^{-\mathrm{i}h_{L-1}\Delta t} + O(\Delta t^2). \tag{1.29}$$

The h_i are the local Hamiltonians acting on bonds i; in general only odd and even bond Hamiltonians will commute in their groups, giving rise to an error in the decomposition. The idea is now simply to use finite-system DMRG in the two-site setup: at each step, one carries out the local infinitesimal time evolution exactly on the two adjacent local sites. This will lead to a new state, a new Schmidt decomposition carried is out in which the system is cut between the two local sites, as before, leading to a new truncation and new reduced basis transformations (2 matrices A adjacent to this bond), which are the choice optimally representing the new state. By doing this for all bonds, one infinitesimal time step is completed.

To do this, one needs the wave function $|\psi\rangle$ in a two-block two-site configuration such that the bond that is currently updated consists of the two free sites. This implies that $|\psi\rangle$ has to be transformed between different configurations. As mentioned above, in finite-system DMRG such a transformation, which was first implemented by White [18] ("state prediction") is routinely used to predict the outcome of large sparse matrix diagonalizations, which no longer occur during time evolution. Here, it merely serves as a basis transformation.

The adaptive time-dependent DMRG algorithm which incorporates the TEBD simulation algorithm in the DMRG framework is therefore set up as follows:

1. Set up a conventional finite-system DMRG algorithm with state prediction using the Hamiltonian at time $t = 0$, $\hat{H}(0)$, to determine the ground state of some system of length L using effective block Hilbert spaces of dimension M. At the end of this stage of the algorithm, we have for blocks of all sizes l reduced orthonormal bases spanned by states $|m_l\rangle$, which are characterized by good quantum numbers. Also, we have all reduced basis transformations, corresponding to the matrices A.

2. For each Trotter time step, use the finite-system DMRG algorithm to run one sweep with the following modifications:

   (i) For each even bond apply the local time evolution $\hat{U}$ at the bond formed by the free sites to $|\psi\rangle$. This is a very fast operation compared to determining the ground state, which is usually done instead in the finite-system algorithm.
   (ii) As always, perform a DMRG truncation at each step of the finite-system algorithm, hence $O(L)$ times.
   (iii) Use White's prediction method to get the representation of the time-evolved state in the setup with the free sites shifted by one.

3. In the reverse direction, apply step (i) to all odd bonds.
4. As in standard finite-system DMRG evaluate operators when desired at the end of some time steps. Note that there is no need to generate these operators at all those time steps where no operator evaluation is desired, which will, due to the small Trotter time step, be the overwhelming majority of steps.

Note that one can also perform every bond evolution operator at each half-sweep, in order. This does not worsen the Trotter error, since in the reverse sweep the operators are applied in reverse order.

The calculation time of adaptive time-dependent DMRG scales linearly in L, as opposed to the static time-dependent DMRG which does not depend on L. The diagonalization of the density matrices (Schmidt decomposition) scales as N^3M^3; the preparation of the local time evolution operator as N^6, but this may have to be done only rarely, e.g., for discontinuous changes of interaction parameters. Carrying out the local time evolution scales as N^4M^2; the basis transformation scales as N^2M^3. As $M \gg N$ typically, the algorithm is of order $O(LN^3M^3)$ at each time step.

The performance of this method has been tested in various applications in the context of ultracold atom physics [8, 32–34], but also for far-from-equilibrium dynamics [35] and for spectral functions [9]; some of these applications will serve as examples in the following.

Before we move on, it should be mentioned that this method for time evolution, while very fast, has weaknesses due to its usage of the Trotter decomposition: first and not so important, there is the Trotter decomposition error depending on the time step. The Trotter error is small and can be reduced to negligible levels by using higher order Trotter decompositions—we are currently using mostly 4th order. More importantly, they are limited to systems with nearest neighbor interactions on a single chain; this problem can be circumvented by introducing larger unit cells such that interactions become nearest neighbor again—given the scaling in N, this is not very feasible. A better approach in such cases is to use the time-step targetted method [10] which, at quite some algorithmic cost in time, does not have these limitations. The main idea is to produce a basis which targets the states needed to represent one small but finite time step. Once this basis is complete enough, the time step is taken and the algorithm proceeds to the next time step. This targetting is intermediate to the other approaches: the Trotter methods target precisely one instant in time at any DMRG step, while Luo et al.'s approach [28] considered the entire range of time to be studied. For the subtle details, I refer to [10].

1.6 Error Analysis: Magnetization Dynamics

In this section, we consider the dynamics of a system far from equilibrium using adaptive time-dependent DMRG [35]. The following example, for which an exact solution is available, shows that time-dependent DMRG can also perform in situations where dynamical DMRG must surely fail. The exact solution allows us to perform DMRG-independent error analysis.

The initial state $|\text{ini}\rangle = |\uparrow \ldots \uparrow\downarrow \ldots \downarrow\rangle$ on a one-dimensional spin-1/2 chain is subjected to the dynamics of the Heisenberg model

$$H = \sum_n S_n^x S_{n+1}^x + S_n^y S_{n+1}^y + J_z S_n^z S_{n+1}^z \equiv \sum_n h_n. \tag{1.30}$$

We set $\hbar = 1$, defining time to be 1/energy with the energy unit chosen as the J_{xy} interaction. The case $J_z = 0$ describes equivalently free fermions on a lattice, and can be solved exactly. In the following we will focus on this case. Note that in that case the initial state with two large ferromagnetic domains separated by a domain wall in the center is a highly excited state; the ground state exhibits power-law decaying antiferromagnetic correlations.

The time evolution delocalizes the domain wall over the entire chain; the magnetization profile for the initial state $|\text{ini}\rangle$ reads [36]:

$$S_z(n,t) = \langle\psi(t)|S_n^z|\psi(t)\rangle = -1/2 \sum_{j=1-n}^{n-1} J_j^2(t), \tag{1.31}$$

where J_j is the Bessel function of the first kind. $n = \ldots, -3, -2, -1, 0, 1, 2, 3, \ldots$ labels chain sites with the convention that the first site in the right half of the chain has label $n = 1$. As the total energy of the system is conserved, the state cannot relax to the ground state. The exact solution reveals a nontrivial behavior with a complicated substructure in the magnetization profile, which is a good benchmark for DMRG.

1.6.1 Possible Errors

Two main sources of error occur in the adaptive t-DMRG:

(i) The *Trotter error* due to the Trotter decomposition. For an nth-order Trotter decomposition [31], the error made in one time step $\mathrm{d}t$ is of order $L\mathrm{d}t^{n+1}$. To reach a given time t one has to perform $t/\mathrm{d}t$ time-steps, such that in the worst case the error grows linearly in time t and the resulting error is of order $L(\mathrm{d}t)^n t$.

(ii) The DMRG *truncation error* due to the representation of the time-evolving quantum state in reduced (albeit "optimally" chosen) Hilbert spaces and to the repeated transformations between different truncated basis sets. While the truncation error ϵ that sets the scale of the error of the wave function and operators is typically very small, here it will strongly accumulate as $O(Lt/\mathrm{d}t)$ truncations are carried out up to time t. This is because the truncated DMRG wave function has norm less than one and is renormalized at each truncation by a factor of $(1-\epsilon)^{-1} > 1$. Truncation errors should therefore accumulate roughly exponentially with an exponent of $\epsilon Lt/\mathrm{d}t$, such that eventually the adaptive t-DMRG will break down at too long times. The accumulated truncation error should decrease considerably with an increasing number of kept DMRG states M. For a fixed time t, it should decrease as the Trotter time step dt is increased, as the number of truncations decreases with the number of time steps $t/\mathrm{d}t$.

At this point, it is worthwhile to mention that our subsequent error analysis should also be pertinent to the very closely related time-evolution algorithm introduced by Verstraete et al. [37], which also involves both Trotter and truncation errors.

We remind the reader that no error is encountered in the application of the local time evolution operator U_n to the state $|\psi\rangle$.

1.6.2 Error Analysis

We use two main measures for the error:

(i) As a measure for the overall error we consider the *magnetization deviation*, the maximum deviation of the local magnetization found by DMRG from the exact result,

$$\mathrm{err}(t) = \max_n |\langle S^z_{n,\mathrm{DMRG}}(t)\rangle - \langle S^z_{n,\mathrm{exact}}(t)\rangle|. \tag{1.32}$$

(ii) As a measure which excludes the Trotter error we use the *forth-back deviation* $FB(t)$, which we define as the deviation between the initial state $|\mathrm{ini}\rangle$ and the state $|fb(t)\rangle = U(-t)U(t)|\mathrm{ini}\rangle$, i.e., the state obtained by evolving $|\mathrm{ini}\rangle$ to some time t and then back to $t=0$ again. If we Trotter-decompose the time evolution operator $U(-t)$ into odd and even bonds in the reverse order of the decomposition of $U(t)$, the identity $U(-t) = U(t)^{-1}$ holds without any Trotter error (up to higher order effects due to the concurrent truncations), and the forth-back deviation has the appealing property to capture the truncation error only.

As the DMRG setup used in this particular calculation did not allow easy access to the fidelity $|\langle \mathrm{ini}|fb(t)\rangle|$ (a calculation which is not a problem in principle, see [34]), the forth-back deviation was defined to be the L_2 measure for the difference of the magnetization profiles of $|\mathrm{ini}\rangle$ and $|fb(t)\rangle$,

$$FB(t) = \left(\sum_n \left(\langle \text{ini}|S_n^z|\text{ini}\rangle - \langle fb(t)|S_n^z|fb(t)\rangle \right)^2 \right)^{1/2} . \tag{1.33}$$

In order to control Trotter and truncation error, two DMRG control parameters are available, the number of DMRG states M and the Trotter time step $\mathrm{d}t$.

The dependence on $\mathrm{d}t$ is twofold: on the one hand, decreasing $\mathrm{d}t$ reduces the Trotter error by some power of $\mathrm{d}t^n$ exactly as in QMC; on the other hand, the number of truncations increases, such that the truncation error is enhanced. It is therefore not a good strategy to choose $\mathrm{d}t$ as small as possible. The truncation error can however be decreased by increasing M.

Consider the dependence of the magnetization deviation err(t) on the number M of DMRG states. In Fig. 1.5, err(t) is plotted for a fixed Trotter time step $\mathrm{d}t = 0.05$ and different values of M. One sees that a M-dependent "runaway time" t_R separates two regimes: for $t < t_R$ (regime A), the deviation grows essentially linearly in time and is independent of M, for $t > t_R$ (regime B), it suddenly starts to grow more rapidly than any power-law as expected of the truncation error. In the inset of Fig. 1.5, t_R is seen to increase roughly linearly with growing M. As $M \to \infty$ corresponds to the complete absence of the truncation error, the M-independent bottom curve of Fig. 1.5 is a measure for the deviation due to the Trotter error alone

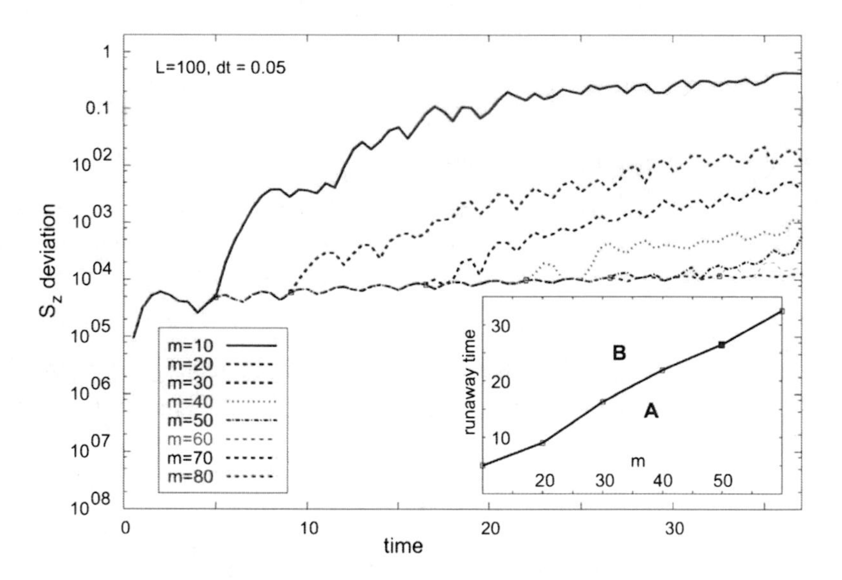

Fig. 1.5 Magnetization deviation err(t) as a function of time for different numbers M of DMRG states. The Trotter time interval is fixed at $\mathrm{d}t = 0.05$. Again, two regimes can be distinguished: For early times, for which the Trotter error dominates, the error is slowly growing (essentially linearly) and independent of M (regime A); for later times, the error is entirely given by the truncation error, which is M-dependent and growing fast (almost exponential up to some saturation; regime B). The transition between the two regimes occurs at a well-defined "runaway time" t_R (small squares). The inset shows a monotonic, roughly linear dependence of t_R on M. From [35]

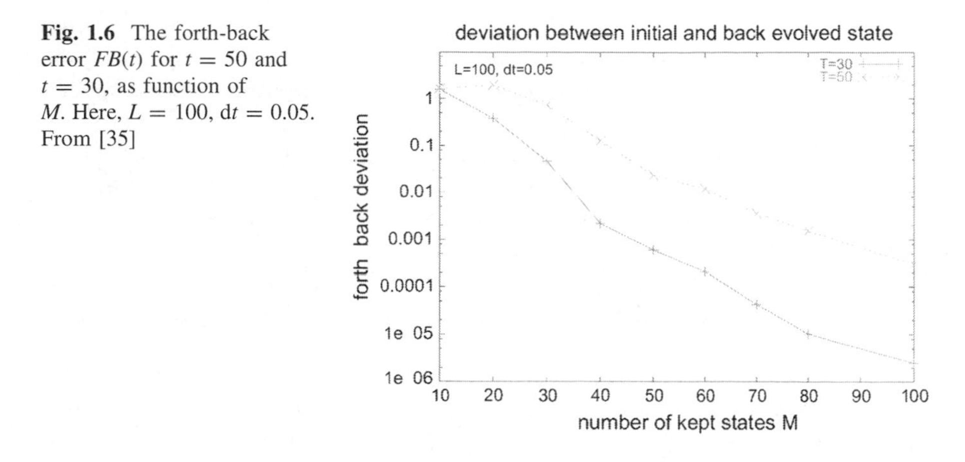

Fig. 1.6 The forth-back error $FB(t)$ for $t = 50$ and $t = 30$, as function of M. Here, $L = 100$, $dt = 0.05$. From [35]

and the runaway time can be read off very precisely as the moment in time when the truncation error starts to dominate.

That the crossover from a dominating Trotter error at short times and a dominating truncation error at long times is so sharp may seem surprising at first, but can be explained easily by observing that the Trotter error grows only linearly in time, but the accumulated truncation error grows almost exponentially in time.

To see that nothing special is happening at t_R, consider also Fig. 1.6, where the Trotter-error free $FB(t)$ is plotted as a function of M, for $t = 30$ and $t = 50$. An approximately exponential increase of the accuracy of the method with growing M is observed for a fixed time. Our numerical results that indicate a roughly linear time-dependence of t_R on M (inset of Fig. 1.5) are the consequence of some balancing of very fast growth of precision with M and decay of precision with t.

The runaway time thus indicates an imminent breakdown of the method and is a good, albeit very conservative measure of available simulation times. We expect the above error analysis for the adaptive t-DMRG to be generic for other models. The truncation error will remain also in approaches that dispose of the Trotter error; maximally reachable simulation times should therefore be roughly similar. Even if for high precision calculation the Trotter error may dominate for a long time, in the long run it is always the truncation error that causes the breakdown of the method at some point in time.

1.7 Spin-Charge Separation in Ultracold Atoms

Following the seminal work of Haldane [38, 39], it has been understood that the low-energy behaviour of 1D quantum liquids is universally described by the Luttinger liquid (LL) picture [40, 41]. A remarkable prediction is spin-charge separation for Fermions: at low energy the excitations of charge and spin completely decouple and propagate with different velocities. The first unequivocal observation was obtained in experiments on the tunneling between two quantum wires [42]. A drawback of condensed-matter setups such as this is that the microscopic interactions strongly influence spin-charge separation, but are neither tunable nor known to some precision. Here, as in other condensed matter problems, the fact that in ultracold gases in optical lattices strong correlations can be studied with unprecedented control and tunability of the parameters might turn out to be very helpful. In fact, an 'atomic quantum wire' configuration in an array of thousands of parallel atom waveguides was realized in ultracold Fermi gases by the application of a strong two dimensional optical lattice [43]. Previous proposals to use cold atoms for studying spin-charge separation [44, 45] were limited by necessary analytical approximations, which do not hold for the strong and localized perturbations that would have to be created experimentally in current, relatively small systems. For a quantitative description of spin-charge separation one needs a microscopic description, which here is given almost perfectly by the Hubbard model. It is an essential new feature of cold atoms in optical lattices that parameters can be changed dynamically and the resulting time evolution can be studied. This gives direct access to the real-time dynamics of strongly correlated systems, an ideal testbed for real-time DMRG: numerical results of the real-time dynamics of a 1D Hubbard model of up to 128 sites can be obtained easily [33], far beyond previous possibilities [46].

We start from the standard Hubbard model

$$
\begin{aligned}
H = -J \sum_{j,\sigma} \left(c^{\dagger}_{j+1,\sigma} c_{j,\sigma} + h.c. \right) + U \sum_{j} n_{j,\uparrow} n_{j,\downarrow} \\
+ \sum_{j,\sigma} \varepsilon_{j,\sigma} \hat{n}_{j,\sigma}
\end{aligned} \tag{1.34}
$$

for Fermions in 1D, where we call the hopping matrix element J, to avoid confusion with time t. Setting $J = 1$ and $\hbar = 1$ time is measured in units of $\hbar/J = 1$. Special for the cold atom setup is a spin-dependent local on-site energy $\varepsilon_{j,\sigma}$, describing both a possible smooth harmonic confinement potential and time-dependent local potentials which allow to perturb the system. One introduces a 'charge' density $n_{\mathrm{c}} = n_{\uparrow} + n_{\downarrow}$ and a 'spin' density $n_{\mathrm{s}} = n_{\uparrow} - n_{\downarrow}$; in a realization with cold gases, where the spin degrees of freedom are represented by two different hyperfine levels, and 'charge' density is particle density. The ratio $u = U/J$ can easily be changed experimentally by varying the depth of the optical lattice.

Experimentally, the density perturbations may be generated by a blue- or red-detuned laser beam tightly focused perpendicular to an array of atomic wires, which generates locally repulsive or attractive potentials for the atoms in the wires. In practice, the perturbations due to an external laser field are quite strong, typically of the order of the recoil energy E_r and thus clearly require a nonperturbative treatment.

In all following calculations system length was up to $L = 128$ sites: several hundred DMRG states were kept. DMRG error analysis reveals that all density distributions shown are exact for all practical purposes, with controlled errors of less than $O(10^{-3})$.

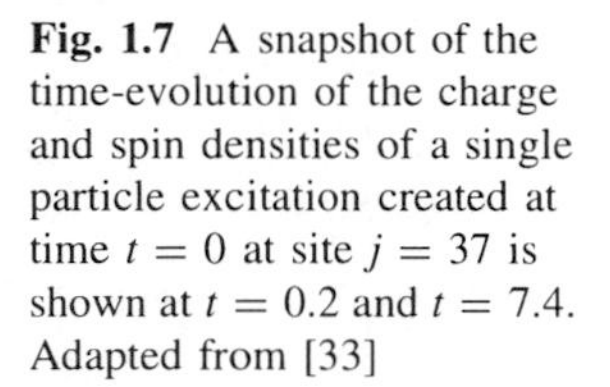

Fig. 1.7 A snapshot of the time-evolution of the charge and spin densities of a single particle excitation created at time $t = 0$ at site $j = 37$ is shown at $t = 0.2$ and $t = 7.4$. Adapted from [33]

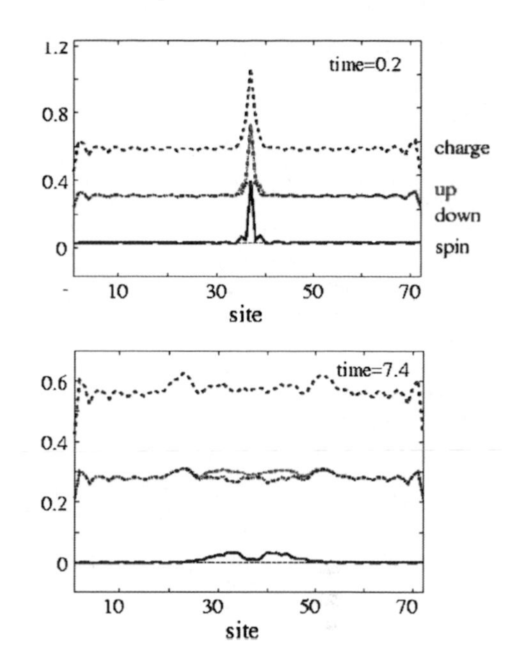

In order to study the behavior of a single particle excitation, the time evolution of the system with one additional particle with spin up added at time $t = 0$ on site j to the ground state, was calculated numerically. In Fig. 1.7 a snapshot of the resulting evolution of the densities is shown for time $t = 7.4$. Remarkably, even after a such a short time separate wave packets in spin and charge can be seen.

In ultracold atom experiments, adding a single particle is not possible. Rather, spin-specific density perturbations can be created as discussed above. We therefore start with a homogeneous system which is perturbed by a potential $\varepsilon_{j,\uparrow}$ localized at the chain center which couples only to the $\uparrow$-Fermions, i.e.,

$$\varepsilon_{j,\uparrow}(t) \propto \exp\{-[j-(L-1)/2]^2/8\}\theta(-t). \tag{1.35}$$

The potential is assumed to have been switched on slowly enough for equilibration, and is then switched off suddenly at time $t = 0$. In Fig. 1.8a the density distribution of the state at an early time is shown as obtained by DMRG. The external potential (1.35) generates a dominant perturbation in the $\uparrow$-Fermion distribution by direct coupling and, indirectly, a smaller perturbation in the $\downarrow$-density due to the repulsive interaction between the different spin species. The wave packets in $\uparrow$ and $\downarrow$-density hence perform a complicated time evolution (Fig. 1.8). In contrast, the perturbations in the spin and charge density split into two wave packets each moving outwards. Their respective velocities are found to be different as indicated by the arrows in Fig. 1.8b, separating spin and charge.

In the limit of an infinitesimal perturbation much broader than the average interparticle spacing, both spin and charge velocities are known analytically from the Bethe ansatz [47–50]. To compare our numerical findings to the exact charge

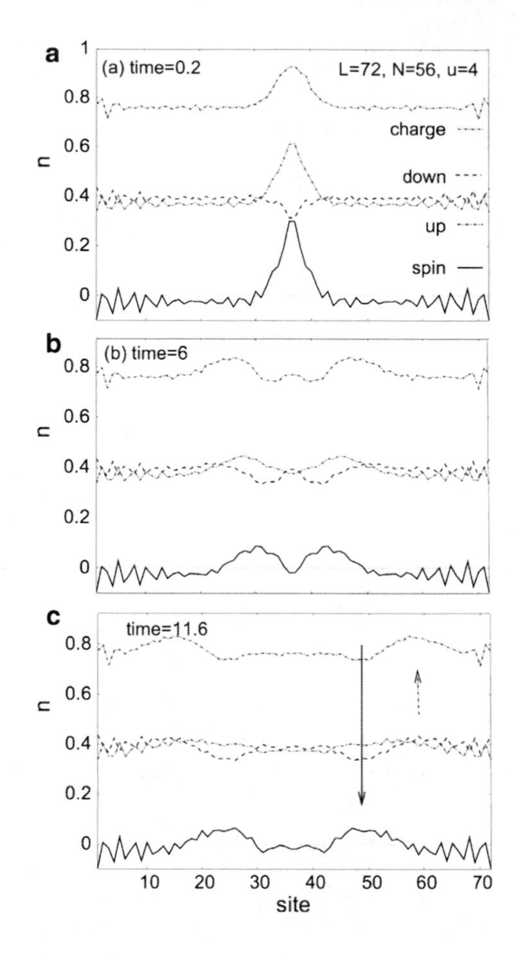

Fig. 1.8 Snapshots of the evolution of the density distribution are shown at different times. At $t = 0$, a wave packet is present in the center of the system in both the spin and the charge density. Each of these splits up into two packets which move with the same velocity in opposite directions. The velocity of the charge wave and the spin wave are different. $u = U/J = 4$, background density is $n_0 = 0.78$. Cf. [33]

velocity, we create pure charge density perturbations, by applying the potential of (1.35) to both species, i.e., $\varepsilon_{j,\uparrow} = \varepsilon_{j,\downarrow}$, and calculate their time-evolution after switching off the potential. The charge velocity is determined from the propagation of the maximum (minimum) of the charge density perturbation for bright (amplitude $\eta_c > 0$) and gray ($\eta_c < 0$) perturbations, respectively. In Fig. 1.9 the charge velocities for various background densities n_0 and perturbation amplitudes η_c are shown. We find good agreement, if we plot the charge velocity versus the charge density at the maximum (minimum), i.e., $n_c = n_0 + \eta_c$. The velocity of the maximum (minimum) of the wave packet is therefore determined by the value of the charge density at the maximum (minimum), not the background density.

This stays true even for strong perturbations $\eta_c \approx \pm 0.1$ which corresponds to 20% of the charge density. The charge velocity is thus robust against separate changes of the background density n_0 and the height of the perturbation η_c.

The previous results indicate that adaptive time-dependent DMRG is a highly performing method, able to answer complicated questions of dynamics in strongly correlated systems. To actually connect to experiments in the context of ultracold

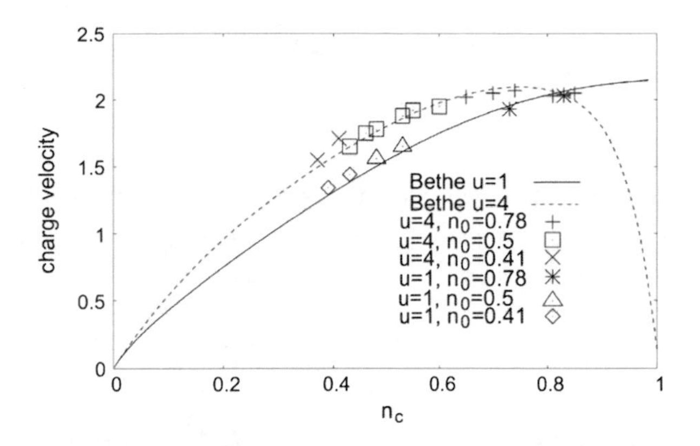

Fig. 1.9 Exact results for the charge velocity obtained by the Bethe ansatz are (lines) compared to the numerical results of the adaptive t-DMRG. The numerical results correspond to different heights of the perturbations at various charge background densities n_0. n_c is the charge density at the maximum/minimum of the charge density perturbation. The uncertainties are of the order of the size of the symbols and stem mainly from the determination of the velocity. From [33]

atoms, one has to take into account also the harmonic trapping potential, which may lead to a coexistence between a liquid (charge conducting) phase at the edges and a Mott-insulating phase in the center of the trap. Also, experiments can only measure quantities averaged over ~ 10 lattice sites. These questions can be addressed in detail by DMRG; as they are not pertinent here, I refer to the literature [33].

1.8 Finite Temperature

After the previous discussion on the difficulties of simulating the time-evolution of pure states in subsets of large Hilbert spaces it may seem that the time-evolution of mixed states (density matrices) is completely out of reach. It is, however, easy to see that a thermal density matrix $\hat{\rho}_\beta \equiv \exp[-\beta \hat{H}]$ can be constructed as a pure state in an enlarged Hilbert space and that Hamiltonian dynamics of the density matrix can be calculated considering just this pure state (dissipative dynamics being more complicated). In the DMRG context, this has first been pointed out by Verstraete et al. [37] and Zwolak and Vidal [51], using essentially information-theoretical language; it has also been used previously in pure statistical physics language in e.g., high-temperature series expansions [52].

To this end, consider the completely mixed state $\hat{\rho}_0 \equiv 1$. Let us assume that the dimension of the local physical state space $\{|\sigma_i\rangle\}$ of a physical site is N. Introduce now a local auxiliary state space $\{|\tau_i\rangle\}$ of the same dimension N on an auxiliary site. The local physical site is thus replaced by a rung of two sites, and a one-dimensional chain by a two-leg ladder of physical and auxiliary sites on top and bottom rungs. Prepare now each rung i in the Bell state

$$|\psi_0^i\rangle = \frac{1}{\sqrt{N}}\left[\sum_{\sigma_i=\tau_i}^{N} |\sigma_i\tau_i\rangle\right]. \tag{1.36}$$

Other choices of $|\psi_0^i\rangle$ are equally feasible, as long as they maintain in their product states maximal entanglement between physical states $|\sigma_i\rangle$ and auxiliary states $|\tau_i\rangle$. Evaluating now the expectation value of some local operator $\hat{O}_\sigma^i$ acting on the physical state space with respect to $|\psi_0^i\rangle$, one finds

$$\langle\psi_0^i|\hat{O}_\sigma^i|\psi_0^i\rangle = \sum_{\sigma_i=\tau_i}\sum_{\sigma_i'=\tau_i'}\frac{1}{N}\left[\langle\sigma_i\tau_i|\hat{O}_\sigma^i \otimes 1_\tau^i|\sigma_i'\tau_i'\rangle\right].$$

The double sum collapses to

$$\langle\psi_0^i|\hat{O}_\sigma^i|\psi_0^i\rangle = \frac{1}{N}\sum_{\sigma_i=1}^{n}\langle\sigma_i|\hat{O}_\sigma^i|\sigma_i\rangle,$$

and we see that the expectation value of $\hat{O}_\sigma^i$ with respect to the pure state $|\psi_0^i\rangle$ living on the product of physical and auxiliary space is identical to the expectation value of $\hat{O}_\sigma^i$ with respect to the completely mixed local physical state, or

$$\langle\hat{O}_\sigma^i\rangle = \mathrm{Tr}_\sigma\hat{\rho}_0^i\hat{O}_\sigma^i \tag{1.37}$$

where

$$\hat{\rho}_0^i = \mathrm{Tr}_\tau|\psi_0^i\rangle\langle\psi_0^i|. \tag{1.38}$$

This generalizes from rung to ladder using the density operator

$$\hat{\rho}_0 = \mathrm{Tr}_\tau|\psi_0\rangle\langle\psi_0|, \tag{1.39}$$

where

$$|\psi_0\rangle = \prod_{i=1}^{L}|\psi_0^i\rangle \tag{1.40}$$

is the product of all local Bell states, and the conversion from ficticious pure state to physical mixed state is achieved by tracing out all auxiliary degrees of freedom.

At finite temperatures $\beta > 0$ one uses

$$\hat{\rho}_\beta = e^{-\beta\hat{H}/2}\cdot 1\cdot e^{-\beta\hat{H}/2} = \mathrm{Tr}_\tau e^{-\beta\hat{H}/2}|\psi_0\rangle\langle\psi_0|e^{-\beta\hat{H}/2},$$

where we have used (1.39) and the observation that the trace can be pulled out as it acts on the auxiliary space and $e^{-\beta\hat{H}/2}$ on the physical space. Hence,

$$\hat{\rho}_\beta = \mathrm{Tr}_\tau|\psi_\beta\rangle\langle\psi_\beta|, \tag{1.41}$$

where $|\psi_\beta\rangle = e^{-\beta\hat{H}/2}|\psi_0\rangle$. Similarly, this finite-temperature density matrix can now be evolved in time by considering $|\psi_\beta(t)\rangle = e^{-i\hat{H}t}|\psi_\beta(0)\rangle$ and $\hat{\rho}_\beta(t) = \mathrm{Tr}_\tau|\psi_\beta(t)\rangle\langle\psi_\beta(t)|$. The calculation of the finite-temperature time-dependent properties of, say, a Hubbard chain, therefore corresponds to the imaginary-time and real-time evolution of a Hubbard ladder prepared to be in a product of special rung states. Time evolutions generated by Hamiltonians act on the physical leg of the ladder only. As for the evaluation of expectation values both local and auxiliary degrees of freedom are traced on the same footing, the distinction can be completely dropped but for the time-evolution itself. Code-reusage is thus almost trivial. Note also that the initial infinite-temperature pure state needs only $M = 1$ block states to be described exactly in DMRG as it is a product state of single local states. Imaginary-time evolution (lowering the temperature) will introduce entanglement such that to maintain some desired DMRG precision M will have to be increased.

References

1. Greiner, M., Mandel, O., Esslinger, T., Hänsch, T.W., Bloch, I.: Nature (London) **415**, 39 (2002)
2. Köhl, M., Moritz, H., Stöferle, T., Günter, K., Esslinger, T.: Phys. Rev. Lett. **94**, 080403 (2005)
3. White, S.R.: Phys. Rev. Lett. **69**, 2863 (1992)
4. White, S.R.: Phys. Rev. B **48**, 10345 (1993)
5. Schollwöck, U.: Rev. Mod. Phys. **77**, 259 (2005)
6. Peschel, I., et al. (eds.): Density-Matrix Renormalization. Springer, Berlin (1999)
7. Vidal, G.: Phys. Rev. Lett. **93**, 040502 (2004)
8. Daley, A.J., Kollath, C., Schollwöck, U., Vidal, G.: J. Stat. Mech.: Theor. Exp. P04005 (2004)
9. White, S.R., Feiguin, A.: Phys. Rev. Lett. **93**, 076401 (2004)
10. Feiguin, A., White, S.R.: Phys. Rev. B **72**, 020404 (2005)
11. Fannes, M., Nachtergaele, B., Werner, R.F.: Europhys. Lett. **10**, 633 (1989)
12. Fannes, M., Nachtergaele, B., Werner, R.F.: Comm. Math. Phys. **144**, 3 (1992)
13. Klümper, A., Schadschneider, A., Zittartz, J.: Europhys. Lett. **24**, 293 (1993)
14. Östlund, S., Rommer, S.: Phys. Rev. Lett. **75**, 3537 (1995)
15. Dukelsky, J., Martin-Delgado, M.A., Nishino, T., Sierra, G.: Europhys. Lett. **43**, 457 (1998)
16. Takasaki, H., Hikihara, T., Nishino, T.: J. Phys. Soc. Jpn. **68**, 1537 (1999)
17. Verstraete, F., Porras, D., Cirac, J.I.: Phys. Rev. Lett. **93**, 227205 (2004)
18. White, S.R.: Phys. Rev. Lett. **77**, 3633 (1996)
19. White, S.R.: Phys. Rev. B **72**, 180403 (2005)
20. Latorre, J.I., Rico, E., Vidal, G.: Quantum Inf. Comut. **4**, 48 (2004)
21. Gaite, J.: quant-ph/0301120
22. Callan, C., Wilczek, F.: Phys. Lett. **333**, 55 (1994)
23. Gaite, J.:Mod. Phys. Lett. A **16**, 1109 (2001)
24. Verstraete, F., Cirac, J.I.: cond-mat/0407066
25. Kühner, T., White, S.R.: Phys. Rev. B **60**, 335 (1999)
26. Jeckelmann, E.: Phys. Rev. B **66**, 045114 (2002)
27. Cazalilla, M., Marston, B.: Phys. Rev. Lett. **88**, 256403 (2002)

28. Luo, H.G., Xiang, T., Wang, X.Q.: Phys. Rev. Lett. **91**, 049701 (2003)
29. Schmitteckert, P.: Phys. Rev. B **70**, 121302 (2004)
30. Vidal, G.: Phys. Rev. Lett. **91**, 147902 (2003)
31. Suzuki M.: Prog. Theor. Phys. **56**, 1454 (1976)
32. Kollath, C., Schollwöck, U., von Delft, J., Zwerger, W.: Phys. Rev. A **71**, 053606 (2005)
33. Kollath, C., Schollwöck, U., Zwerger, W.: Phys. Rev. Lett. **95**, 176401 (2005)
34. Trebst, S., Schollwöck, U., Troyer, M., Zoller, P.: Phys. Rev. Lett. **96**, 250402 (2006)
35. Gobert, D., Kollath, C., Schollwöck, U., Schütz, G.: Phys. Rev. E **71**, 036102 (2005)
36. Antal, T., Racz, Z., Rakos, A., Schütz, G.: Phys. Rev. E **59**, 4912 (1999)
37. Verstraete, F., Garcia-Rípoll, J.J., Cirac, J.I.: Phys. Rev. Lett. **93**, 207204 (2004)
38. Haldane, F.D.M.: J. Phys. C: Solid State Phys. **14**, 2585 (1981)
39. Haldane, F.D.M.: Phys. Rev. Lett. **47**, 1840 (1981)
40. Voit J.: Rep. Prog. Phys. **58**, 977 (1995)
41. Giamarchi, T.: Quantum Physics in One Dimension. Oxford University press, New York (2004)
42. Auslaender, O.M., et al.: Science **308**, 88 (2005)
43. Moritz, H., Stöferle, Th., Günter, K., Köhl, M., Esslinger, T.: Phys. Rev. Lett. **94**, 210401 (2005)
44. Recati, A., Fedichev, P.O., Zwerger, W., Zoller, P.: Phys. Rev. Lett. **90**, 020401 (2003)
45. Kecke, L., Grabert, H., Häusler, W.: Phys. Rev. Lett. **94**, 176802 (2005)
46. Hallberg, K., Aligia, A.A., Kampf, A.P., Normand, B.: Phys. Rev. Lett. **93**, 067203 (2004)
47. Lieb, E.H., Wu, F.Y.: Phys. Rev. Lett. **20**, 1445 (1968)
48. Shiba, H.: Phys. Rev. B **6**, 930 (1972)
49. Coll, C.F.: Phys. Rev. B **9**, 2150 (1974)
50. Schulz, H.J.: Phys. Rev. Lett. **64**, 2831 (1990)
51. Zwolak, M., Vidal, G.: Phys. Rev. Lett. **93**, 207205 (2004)
52. Bühler, A., Elstner, N., Uhrig, G.S.: Eur. Phys. J. B **16**, 475 (2000)

Chapter 2
Molecular Conductance from Ab Initio Calculations: Self Energies and Absorbing Boundary Conditions

Ferdinand Evers and Andreas Arnold

2.1 Introduction

The most important driving force in the research field of *Molecular Electronics* are prospects on technological applications—whence the name—in entirely new realms of system parameters [1]. The development of these new technologies also requires serious progress in several disciplines of fundamental sciences including both, theory and experiment. One of the major theoretical challenges is the quantitative description of transport through a molecule with a given contact geometry [2–6].

In order to appreciate the caliber of the problem, recall that describing transport requires to keep track of two aspects of physical reality, each by itself posing a task of considerable difficulty. Needed are (a) a good knowledge of molecular states, i.e., of energy levels and orbitals, which is not easy to obtain, since they experience a strong influence by Coulomb interactions on the molecule, and (b) a thorough understanding of the hybridization of these orbitals with the electronic lead states, so as to predict the broadening, i.e., the "life time", of molecular energy levels. This seriously complicates matters for ab initio calculations, because inevitably a macroscopic number of degrees of freedom is involved. We are facing here a classical dilemma: each of the two problems—interactions on the molecule and the macroscopic number of lead atoms—by itself can be dealt with reasonably well, but only at the expense of applying methods that exclude a simultaneous solution

F. Evers (✉)
Institute of Nanotechnology, Karlsruhe Institute of Technology,
76021 Karlsruhe, Germany
e-mail: Evers@int.fzk.de

A. Arnold
Institut für Theorie der Kondensierten Materie, Karlsruhe Institute of Technology,
76128 Karlsruhe, Germany

M. Vojta et al. (eds.), *CFN Lectures on Functional Nanostructures – Volume 2*,
Lecture Notes in Physics 820, DOI: 10.1007/978-3-642-14376-2_2,

of the other problem. In a sense, we find ourselves in a situation not unlike Ulysses, when he was trying to pass by Scylla and Charybdis [7].

In this paper we present a method, that simplifies (b), i.e., including macroscopic electrodes into ab initio calculations. The incarnation that we put forward in this communication, operates in those instances where the calculation of transport coefficients builds upon a formalism in terms of Green's functions. The basic idea developed here, however, is much more general and may also be of use for example in transport calculations based on the density matrix renormalization group [8].

A typical example of a Green's function based transport theory met in cases where the quasi-particles are effectively non-interacting, is the Landauer–Büttiker approach to transport [9, 10]. Here, the conductance (in units e^2/h) is expressed as a transmission at the Fermi energy, $g = T(E_F)$. Explicitly, $T(E)$ has a representation

$$T(E) = \mathrm{Tr}\,(G\Gamma_\ell G^\dagger \Gamma_r) \tag{2.1}$$

which may be derived using elementary scattering theory [3, 11, 12], the Keldysh technique [13] or the Kubo formula [14, 15], in principle.

The "dressed" Green's functions, G, required in any of these approaches describe the propagation of particles with energy E on the molecule in the presence of the electrodes. The external leads, left and right, enter these functions by self energy contributions, $\Sigma_{\ell,r}$, one for every electrode. They relate G to the Green's function of the isolated molecule, $G_\mathcal{M}$ by

$$G^{-1}(E) = G_\mathcal{M}^{-1}(E) - \Sigma_\mathcal{M}(E) \tag{2.2}$$

and include all the effects of coupling to the left and right leads, $\Sigma_\mathcal{M} = \Sigma_\ell + \Sigma_r$. Also, they determine the level broadening $\Gamma_{\ell,r} = i(\Sigma_{\ell,r} - \Sigma_{\ell,r}^\dagger)$ appearing in (2.2). Equation 2.2 should be understood as a family of matrix equations with resolvent operators $G, G_\mathcal{M}$, parameterized by energy, E; Green's functions are actually the matrix elements of these operators $G(E, \mathbf{x}, \mathbf{x}') = \langle \mathbf{x}|G(E)|\mathbf{x}'\rangle$ and $G_\mathcal{M}(E, \mathbf{x}, \mathbf{x}') = \langle \mathbf{x}|G_\mathcal{M}(E)|\mathbf{x}'\rangle$.

The calculation of the exact couplings, $\Sigma_{\ell,r}$, is usually fairly troublesome. In the simplest case, when the electron interaction can be appropriately dealt with by an effective single-particle model, the couplings take a structure

$$\Sigma_\chi = t_\chi G_{S;\chi} t_\chi^\dagger, \quad \chi = \ell, r \tag{2.3}$$

where $t_{\ell,r}$ denote the two hopping matrices that connect the molecular junction with the left and right electrodes [13]. The "surface" Green's function, $G_S; \ell, r$, introduced here describes the propagation of quasi-particles on the electrodes in the presence of the contact surface. Even in this situation calculating $\Sigma_\mathcal{M}$ is not really easy. Complications arise since (a) $\Sigma_\mathcal{M}$ should include *macroscopic* leads plus contact geometry and (b) the hopping matrix t is (normally) not just of the nearest neighbor type, so the contact surface also involves sub-surface layers of electrode atoms, in general.

The procedure to be proposed in this communication simplifies the conductance calculation by essentially eliminating the step of evaluating $\Sigma_{\mathcal{M}}$. It works after the molecule has been redefined. The "extended" molecule, $e\mathcal{M}$, not only includes the original molecule but also pieces of the left and right electrodes:

$$G^{-1}(E) = G_{e\mathcal{M}}^{-1} - \Sigma_{e\mathcal{M}}. \tag{2.4}$$

Our key observation is the following: while the molecular conductance is crucially dependent on microscopic details incorporated in $\Sigma_{\mathcal{M}}$ it is completely indifferent towards details in $\Sigma_{e\mathcal{M}}$, *if* $e\mathcal{M}$ includes sufficiently many electrode atoms. As a consequence, there is no need to use the exact self energy $\Sigma_{e\mathcal{M}}$ in order to obtain (in principle) exact results.[1] One can replace it by a simplistic model coupling of the type

$$\langle \mathbf{x} | \, \Sigma_{e\mathcal{M}}(E) \, | \mathbf{x}' \rangle \rightarrow i\eta(\mathbf{x}) \, \delta(\mathbf{x} - \mathbf{x}'), \tag{2.5}$$

where we have introduced a "local leakage function" $\eta(\mathbf{x})$. It is crucial to our method, that fine tuning $\eta(\mathbf{x})$ is obsolete once certain criteria to be specified in Sect. 2.3 are satisfied.

The outline of this paper is as follows. In Sect. 2.2 we recapitulate in broad terms the physical effects encoded in the self energy formalism. The concept of the extended molecule will emerge quite naturally from these considerations. They will also illuminate under which conditions (2.5) can be justified. In the following Sects. 2.3 and 2.4 we will present a series of model problems. In order to demonstrate the principle, we begin in Sect. 2.3 with a tight binding chain, for which numerical results can be compared against analytical solutions. To illustrate the usefulness in practically relevant cases, the conductance of di-thiophenyl is investigated in Sect. 2.4 using an approach to transport based on the density functional theory and the quantum chemistry package TURBOMOLE [16–18]. This study is also intended to reveal the limits of the ansatz (2.5).

2.2 Basic Physics of Self Energies

In this section we will give a more precise definition and a justification of the procedure (2.5) for constructing a self energy, which is based on physical

[1] The words "exact result" have in the present context a restricted meaning: they refer to the exact solution of the single-particle scattering problem, which can be stated once the Kohn–Sham orbitals and energies are given. Under which conditions—if at all—scattering theory based on (ground state) Kohn–Sham orbitals could give an exact description of the full many body problem, this is an important question which, however, goes well beyond the scope of the present article.

arguments. In order to explain the logic, we first recall on a qualitative level which physical information is carried by the original self energy, $\Sigma_{\mathcal{M}}$. We shall illustrate then, how this information is transferred into $G_{e\mathcal{M}}$ by reformulating the problem to calculate G in terms of an extended molecule. If enough metal atoms have been included in $e\mathcal{M}$, the "information transfer" will be complete. Then the remaining information in the self energy $\Sigma_{e\mathcal{M}}$ of $e\mathcal{M}$ is trivial, i.e., it is no longer molecule specific. Therefore, $\Sigma_{e\mathcal{M}}$ is apt to simple approximations like (2.5).

2.2.1 Self Energy of the Molecule $\Sigma_{\mathcal{M}}$

The self energy, $\Sigma_{\mathcal{M}}$, that appears in (2.2),

$$\Sigma_{\mathcal{M}}(E) = G_{\mathcal{M}}^{-1}(E) - G^{-1}(E)$$

has two qualitatively different effects which are incorporated into its hermitian and anti-hermitian constituents.

2.2.1.1 Hermitian Constituent

The eigenvalues of $G_{e\mathcal{M}}$ for the isolated molecule are real numbers. Due to the hermitian piece, $\delta H_{\mathcal{M}} = (\Sigma_{\mathcal{M}} + \Sigma_{\mathcal{M}}^{\dagger})/2$, these eigenvalues undergo a shift, $\Delta\epsilon_{\nu}$, when the molecule is coupled to the electrodes. In the case of weak electron-electron interaction this simple "renormalization" of excitation energies is all that can happen. However, if the interaction is strong, such that the electrons are highly correlated, additional and qualitatively different effects can occur. A most prominent representative is the Kondo effect, observation of which has been reported in various recent experiments [19–21]. It manifests itself in the spectral function of the coupled molecule

$$A(E) = (i/2\pi)\ \mathrm{Tr}(G(E) - G^{\dagger}(E)), \tag{2.6}$$

which measures the number of molecular excitations with a certain energy, roughly speaking [22]. Kondo physics is signalized by an additional peak in $A(E)$, the "Abrikosov–Suhl" resonance, which sits right at the Fermi energy of the leads [23]. This resonance is a collective phenomenon involving electrons from the leads and the molecule; it cannot be understood as a renormalization of a molecular energy level alone.

Even in the absence of strong correlation effects, the shift of molecular excitation levels, $\Delta\epsilon_{\nu}$, can have very important consequences for the interpretation of experimental findings. The presence of the metal electrodes can help screening the interaction of electrons on the molecule. As a consequence, the energy difference between the lowest unoccupied molecular energy level (LUMO) and the highest occupied level (HOMO) will generally shrink. In the extreme case, where the

LUMO falls below (or the HOMO above) the Fermi energy of the electrodes, charge will flow onto the molecule such that the molecular junction becomes partially polarized even at equilibrium conditions.

2.2.1.2 Anti-Hermitian Constituent: Exponential Time Evolution and Reservoirs

The hermitian piece of $\Sigma_{\mathcal{M}}$ is basically "just" a modification of the (effective) Hamiltonian. By contrast, the anti-hermitian piece of the self energy, $(\Sigma_{\mathcal{M}} - \Sigma_{\mathcal{M}}^{\dagger})/2$, introduces a qualitatively new aspect, because it gives rise to an imaginary component, $i\gamma_v$, of the eigenvalues of G. It gives the molecular levels, v, a finite lifetime reflecting a simple physical fact: an initial excitation, localized at time $t = 0$ on the molecule, can fade away to be absorbed by the leads, ultimately.

Let us discuss how excitations pass away in more detail so as to see why the electrodes and the thermodynamic limit are important ingredients in understanding the self energy. We begin by noting that Green's functions can describe a time evolution of the physical system. Therefore, the relaxation rates, γ_v, also have a straightforward interpretation in time space. Assume that the molecular junction is prepared in an initial state such that the molecule has an excess charge. Then, the rates γ_v describe an exponential decay in time, $\exp(-\gamma_v t)$, exhibited by each contribution to this charge made from a certain molecular level, v.

Now, the exponential dependence proposed here is implied to be valid at all times including in particular the asymptotic regime $t \to \infty$. This means that the charge is really swallowed up by the electrodes, it never returns to the molecule and only for this reason the relaxation process can ever become complete. In other words, the electrodes act like thermodynamic baths or *reservoirs*. They destroy information about the initial state in the sense that the return time of a signal, i.e., electrons, from the reservoirs is infinite.

As usual, a truly diverging time scale can be realized only with infinitely many degrees of freedom; otherwise return paths (e.g., of electrons) exist with an overall weight that is not vanishing. In this infinite-dimensional Hilbert space the time evolution is unitary, of course. The (anti-hermitian part of the) self energy pops up as a consequence of projecting the full time evolution down to the subspace of the molecule, $\mathcal{M}$, which then can no longer remain unitary. The principle encountered here is well known in the general theory of non-equilibrium phenomena [24].

2.2.1.3 Self Energy and Transport

Clearly, the decay rates γ_v must be closely related to transport properties, because they govern the time evolution of charge exchange between molecules and leads. Note, however, that the self energy of the Green's function contains information only about the total loss rate

$$\Sigma_{\mathcal{M}} = \Sigma_{\ell} + \Sigma_r$$

due to leakage. It does not necessarily keep track of the rates $\Sigma_{\ell,r}$ separately that describe the exchange with the individual leads, left or right. This latter piece of information is important for the transport characteristics, as can be seen, e.g., in the Landauer–Büttiker formula (2.1). In general, it cannot be reconstructed from the $G(E, \mathbf{x}, \mathbf{x}')$ alone, without making further assumptions (e.g., that $\langle \mathbf{x}|\Sigma_{\mathcal{M}}|\mathbf{x}'\rangle$ is block-diagonal with the two diagonal entries resembling $\Sigma_{r,\ell}$, separately).

In order to illustrate the significance of the level shifts $\Delta\epsilon_\nu$ and level broadenings γ_ν for the transport problem, we consider now a situation typical of experiments on molecular conductance. We focus on the case of weakly interacting electrons and call $\delta_{\rm hl}$ the energy gap between the HOMO of the isolated molecule, ϵ_h and its LUMO, ϵ_l: $\delta_{\rm hl} = \epsilon_l - \epsilon_h$. In typical transport experiments one has a situation where $\delta_{\rm hl} \gtrsim 1$ eV. At the same time, the experimentally measured values of the conductance, g, of the molecule only very rarely exceed 0.1. Both observations taken together give a strong indication that for this type of experiments the level broadening of HOMO and LUMO, $\gamma_{\rm h,l}$, is well below the level separation, $\gamma_{\rm h,l} \ll \delta_{\rm hl}$. Roughly speaking, the conductance (2.1) will then be given by a superposition of two Lorenzians,

$$g = \sum_{\rm x=h,l} \frac{\gamma_{{\rm x},\ell}\gamma_{{\rm x},r}}{(E_{\rm F} - \epsilon_{\rm x} - \Delta\epsilon_{\rm x})^2 + (\gamma_{{\rm x},\ell} + \gamma_{{\rm x},r})^2/4}. \tag{2.7}$$

with a Fermi energy of the metal, $E_{\rm F}$, situated in between the values of HOMO and LUMO after coupling, $\epsilon_{\rm H,L} = \epsilon_{\rm h,l} + \Delta\epsilon_{\rm h,l}$.

We add a remark regarding uncertainties in theoretical predictions of level positions and their broadenings. Inaccuracies in calculating absolute values of the level positions tend to induce a shift of the transmission curve, but do not normally change their structure—unless molecular levels happen to cross the Fermi energy of the electrodes, of course. Often, the shift is very similar for all energy levels involved, and therefore it can be partially eliminated when the transmission is plotted over $E - E_{\rm F}$.

Inaccuracies in the level broadening are more severe, since their error turns out to be of the order of unity. The value of the conductivity off resonance is determined by $\gamma^2{}_{\rm h,l}$, and so a quantitative calculation of g under these conditions is very difficult. The source of this error and the question how it can be overcome became a very active field of research, recently [2–6].

2.2.1.4 Relation to the Renormalization Group Method

In this section, we describe the physics incorporated in $\Sigma_{\mathcal{M}}$ from the point of view of an hypothetical renormalization group method. This is to say, we investigate how $\Sigma_{\mathcal{M}}$ evolves when we build up the molecular junction gradually step by step, attaching more and more electrode atoms. The idea is roughly in a spirit similar to the density matrix renormalization group [25], also discussed in this book in the chapter by Ulrich Schollwöck. The flow thus induced will be smooth unless the molecule becomes strongly distorted, which could indicate, for example, dissociation or ionization.

In order to illustrate this evolutionary process, we have performed calculations based on the density functional theory (DFT) using the standard functional BP86 [26, 27]. DFT provides us with an effective single-particle Hamiltonian, $H_{N_{\mathfrak{E}}}$, with eigenvalues ϵ_ν and corresponding eigenstates $|\nu\rangle$. Our interest is how the eigenvalues and eigenfunctions change when we include a gradually increasing number of electrode atoms, $N_{\mathfrak{E}}$, in our model system.

The result of this procedure has been depicted in Fig. 2.1 for the case of the molecule di-thiophenyl (See Fig. 2.2 for the detailed atomic structure). Every eigenfunction is represented by a data point (ϵ_ν, A_ν). The integrated amplitude is defined as

$$A_\nu = \mathrm{Tr}_{\mathcal{M}} |\nu\rangle\langle\nu|$$

where the $\mathrm{Tr}_{\mathcal{M}}$ is over the projected segment of the Hilbert space that is associated with the molecular degrees of freedom. Our calculation is performed using a local basis set $|X\ell\rangle$ (TZVPP [28]), with basis functions labeled by atomic positions, X, and orbital quantum numbers, ℓ. When evaluated in this basis, the $\mathrm{Tr}_{\mathcal{M}}$ is a sum over all basis states that belong to atoms of the molecule, excluding lead atoms.

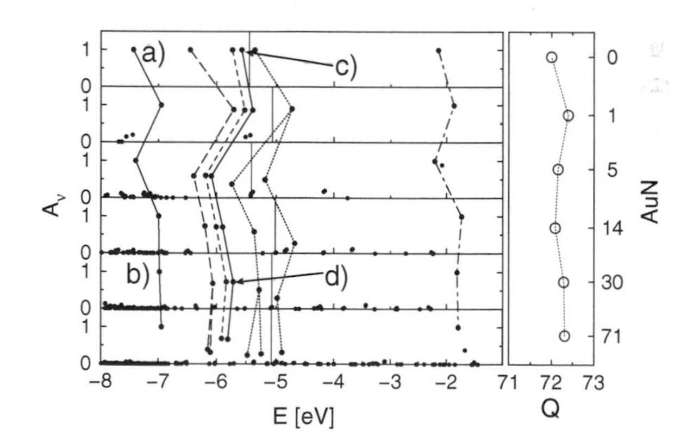

Fig. 2.1 Di-thiophenyl between Au pyramids (symmetric coupling, see Fig. 2.2). *Left panel*: flow of energy of different orbitals with increasing number, $N_{\mathfrak{E}}$, of Au-electrode atoms as indicated at the right-hand axis. (Orbitals at *a*, *b*, *c*, *d* are shown in Fig. 2.3). Each orbital is characterized by its energy and weight A_ν on the molecule. The vertical bars near −5 eV mark the (center of the) HOMO-LUMO gap. The evolution of the six orbitals of the isolated molecule has been indicated by vertical lines. *Right panel*: evolution of charge, Q, accumulated on the molecule (including sulphur atoms)

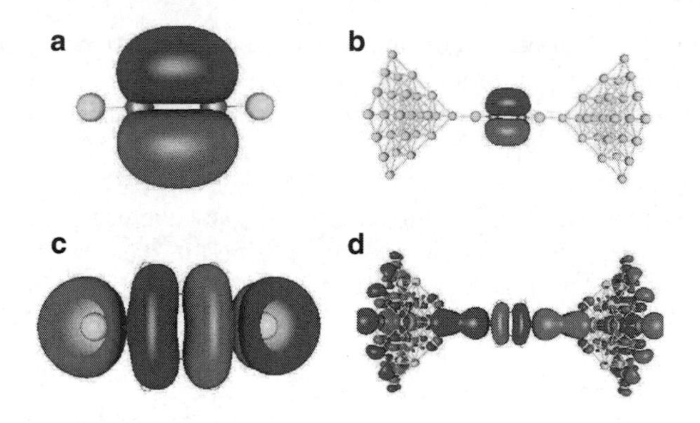

Fig. 2.2 Plots of two orbitals, ψ, for di-thiophenyl without electrodes (*left*) and with $N_{\mathcal{E}} = 30$ Au atoms attached (*right*). Orbital **a** hybridizes only very little with the electrodes and is almost unaffected after coupling, **b** By contrast, orbital **c** becomes completely delocalized in the metal, see **d** *Gray* (*black*) colors indicate regions with ψ positive (negative)

Fig. 2.1 shows how the six orbitals of the molecule in the gas-phase shift and hybridize with $N_{\mathcal{E}}$ increasing from 0, 1, 5, 14, 30, 71. To illustrate this, we have also given the Kohn–Sham wavefunctions of two representative states in Fig. 2.3.

The overall plot clearly shows that the original orbitals survive the coupling to the electrodes and therefore contribute as resonances to the transport characteristics. The initial evolution at small $N_{\mathcal{E}}$ is not very smooth, because (a) attaching the first few Au atoms cannot be considered a very small perturbation to the molecular system and (b) with high symmetry conformations of cluster atoms, e.g., $N_{\mathcal{E}} = 5$, the electrode configuration is particularly stable. These "stability" islands are interesting in themselves but for our present purpose they deliver parasitic side effects, since they make it more difficult to extrapolate the overall flow. When keeping away from exceptional numbers, e.g., by choosing $N_{\mathcal{E}} = 14, 30, 71$, the evolution shows the expected smooth behavior.

We have already mentioned that the smooth evolution of single-particle levels can also be perturbed, if a prominent molecular level happens to cross the Fermi

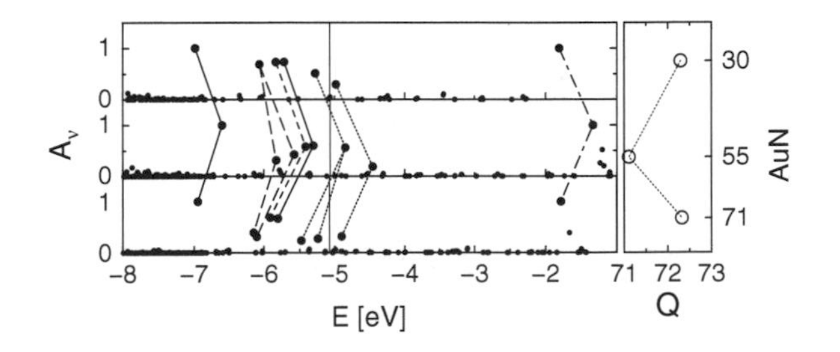

Fig. 2.3 Plot similar to Fig. 2.1 with $N_{\mathcal{E}} = 55$ included. A shift towards higher energies of the prominent orbitals is caused in the *center panel* because of additional, "evanescent" modes appearing

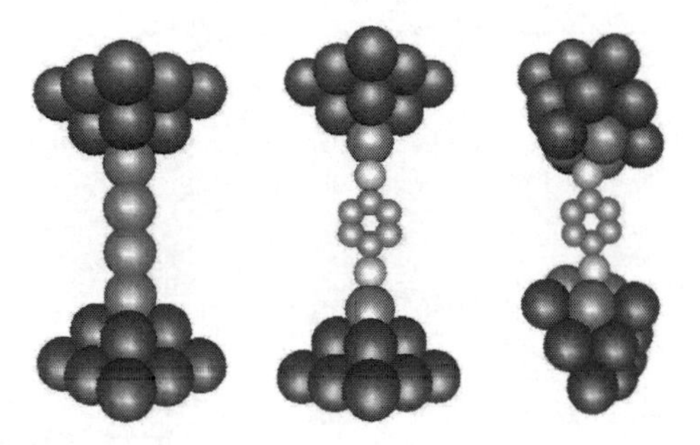

Fig. 2.4 Atomic configurations of molecular junctions attached to pieces of the electrodes. Electrodes are modeled by pyramids, $N_{\mathcal{E}} = 14$ Au atoms each. Dark atoms are surface atoms used in self energy (2.29). *Left*: 2 Au atom wire, $N_{\mathcal{S}} = 13$ (Sect. 4.1). *Center*: di-thiophenyl with (stretched) S–Au_1 coupling; two sulphur, six carbon atoms, $N_{\mathcal{S}} = 13$ (Sect. 4.2). *Right*: di-thiophenyl with relaxed S–Au_3 coupling, $N_{\mathcal{S}} = 11$ (Sect. 4.2)

energy. This can happen in a situation where the HOMO is relatively close to E_F. In such a case small fluctuations of the charge distribution that occur due to the gradual appearance of "evanescent" modes, i.e., invading electrode states with energies in between the prominent molecular orbitals, can lift the (designated) HOMO above E_F at certain electrode configurations.

Such a behavior can be observed for Au_{55} electrodes, as depicted in Fig. 2.4. In this case a molecular state (that turns out to be localized predominantly on the S-atoms) peaks above E_F and therefore is evacuated, leading to a decrease of the charge accumulated on the molecule by $1.2e$. The fate of this prominent mode that has been expelled from the region of occupied energy levels is a fast decay with further increasing $N_{\mathcal{E}}$, because of its relatively strong coupling to the "invaders", see Fig. 2.4.

So far we have witnessed the transformation (or in some cases the decay) of the states that occurs as a consequence of the hybridization of electrode and molecular orbitals. In a sense, this is the analogue of Fermi-liquid theory. The Kondo effect, which in principle could appear for molecular systems that carry a spin, cannot be understood within this picture. This is because the Abrikosov–Suhl resonance is not a shifted molecular level. Instead, it is a *collective phenomenon* and generated by a large number of electrode states. Their energies reside inside the HOMO-LUMO gap of the "dressed" molecule close to the Fermi energy. This effect can be seen with DFT in principle, but the current approximations for the exchange-correlation functional are too crude to capture it. A study with the exact, at present unknown functional should show satellites at E_F produced by lead states that merge with one another if the system size becomes large. If a sufficiently large number of these states are superimposed, a sharp peak, the Abrikosov–Suhl resonance, grows right within the HOMO-LUMO gap. A further typical characteristic of this collective effect is the "pinning" of this resonance at the Fermi energy, irrespective of shifts in the molecular orbital energies that might be induced by a gate, for instance (Fig. 2.5).

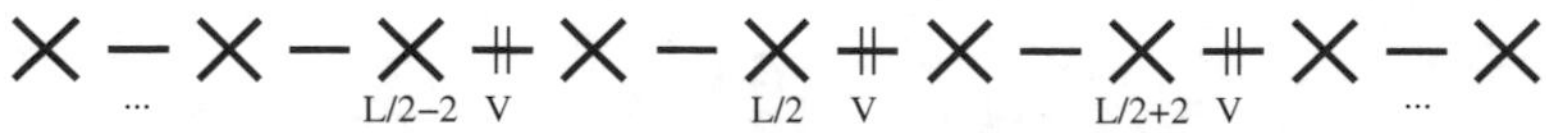

X − X − X ‖ X − X ‖ X − X ‖ X − X

... L/2−2 V L/2 V L/2+2 V ...

Fig. 2.5 Single channel tight binding wire with triple barrier realized by weak links (indicated by *vertical double lines*) as used in model calculations

2.2.2 Extended Molecule and $\Sigma_{e\mathcal{M}}$

The self energy of the original molecule, $\Sigma_{\mathcal{M}}$, can contain a wealth of nontrivial information, it is not a quantity easy to calculate. This was the message of the preceeding section. However, the situation greatly simplifies after the molecule has been redefined. Let us consider an extended molecule, $e\mathcal{M}$, that comprises in addition to the original molecule also a "contact region", i.e., a number $N_{\mathfrak{E}}$ of electrode atoms. The Green's function for an extended system, $G_{e\mathcal{M}}$, is related to the full Green's function G via a new self energy

$$G^{-1}(E) = G^{-1}_{e\mathcal{M}}(E) - \Sigma_{e\mathcal{M}}(E). \tag{2.8}$$

We give two reasons why $\Sigma_{e\mathcal{M}}$ is much easier to handle than $\Sigma_{\mathcal{M}}$.

Imagine the extreme limit, in which far more electrons are located on the metal than on the molecule. Then the HOMO of the big system, ϵ_{H}, is given by the Fermi energy of the metal, E_{F}, up to a small uncertainty which is of the order of the HOMO-LUMO gap of $e\mathcal{M}$, δ_{HL}. In a metal, this gap is inversely proportional to the number of metal electrons in the calculation, $\delta_{\text{HL}} \propto 1/N_{e\mathcal{M}}$, so that the uncertainty of the position of the Fermi energy with respect to $\epsilon_{\text{H,L}}$ can be made arbitrarily small. This is an obvious advantage.

More importantly, the flow of the typical level broadening, $\gamma_{\text{H,L}}$ (related to $\Sigma_{e\mathcal{M}}$), that is driven by increasing the number of electrode atoms, $N_{\mathfrak{E}}$, in the contact region, will lead us into a very tractable regime as we shall see now. It is only a fraction $N_{\mathcal{S}}$ of the $N_{\mathfrak{E}}$ electrode atoms that is actually connected to the rest of the leads, the "outside world". Assume that $N_{\mathfrak{E}}$ grows in such a fashion that the number of these "surface atoms", $N_{\mathcal{S}}$, does not change, so we build a quasi-one-dimensional wire. Then, increasing $N_{\mathfrak{E}}$ implies that the fraction of the wave-function amplitude of extended orbitals located near the surface decays like $N_{\mathcal{S}}/N_{\mathfrak{E}}$, so that $\gamma_{\text{H,L}}$ scales like δ_{HL}. In good metals the ratio of both energies is of the order of the metallic conductance

$$\gamma_{\text{H,L}}/\delta_{\text{HL}} \sim g \gtrsim 1. \tag{2.9}$$

Thereby, the level broadening of HOMO and LUMO of the extended molecule always exceeds their separation if the electrodes are made of a good metal. This is a situation exactly opposite to the problematic one, $\gamma_{\text{h,1}}/\delta_{\text{hl}} \ll 1$, that we have encountered before in the context of (2.7). Summarizing, for the extended molecule, $e\mathcal{M}$, the following hierarchy of inequalities holds

$$\delta_{\mathrm{HL}} \leq \gamma_{\mathrm{H,L}} \ll \gamma_{\mathrm{h,l}} \leq \delta_{\mathrm{hl}}. \tag{2.10}$$

The separation of energy scales implied by (2.10) is the prerequisite for the real gain that one makes when one turns to the extended molecule. The point is that the fine structure in the spectrum of $G_{e\mathcal{M}}$ is of the order of δ_{HL}. The anti-hermitian constituent of $\Sigma_{e\mathcal{M}}$, $\gamma_{\mathrm{H,L}}$, provides the smearing of this fine structure necessary in order to obtain smooth curves, e.g., for spectral and transmission functions. The details of this smearing have very little impact on the resulting curves, as the interesting structures are related to energy scales $\delta_{\mathrm{h,l}}$ and $\gamma_{\mathrm{h,l}}$, which exceed δ_{HL} and $\gamma_{\mathrm{H,L}}$ by a parametrically large factor.

2.2.3 $\Sigma_{e\mathcal{M}}$ and Absorbing Boundary Conditions

There is yet another, perhaps particularly intuitive way to understand the principal difference between $\Sigma_{\mathcal{M}}$ and $\Sigma_{e\mathcal{M}}$. It will serve as a motivation for the proposed approximation (2.5).

Let us assume we opted for an investigation of transport properties in the time domain, e.g., by propagating wave packets. Then, we would study the time evolution of a wave packet, localized at $t = 0$ at some initial position on the molecule. In particular, we can investigate how wave packets leak out of the molecule into the contacts such that they gradually disappear. When performing such an investigation systematically in the presence of leads, one can in principle collect enough information in order to reconstruct the Fourier transform of the retarded Green's function, $G(t)$.

There is a condition on the observation time, T. In order to have an energy resolution $\gamma_{l,r}$ we need $T \gtrsim \gamma_{l,r}^{-1}$. This is troublesome if the contact size maintained in our calculation is not sufficiently large. After some time, the leakage will hit the outside walls of the contact, be reflected back, and finally, after the "dwell" time, τ_D, rearrive at the molecule; hence, we calculate $G_{e\mathcal{M}}(t)$ instead of $G(t)$. The energy resolution that we achieve with such a calculation, cannot exceed h/τ_D. The best that we can hope for is $\tau_{\mathrm{D}} \sim \delta_{\mathrm{HL}}^{-1}$ but only if the contact acts as a fully chaotic cavity. At longer times the signal from the decaying wavepacket will be superimposed by the cavity modes that describe how the wave packet sloshes back and forth inside the electrodes. We shall demonstrate this effect in Sect. 2.3 looking at an explicit example.

The salient point we wish to make is that the minimum dwell time in the cavity should be so long that the wave packet has enough time to evacuate before the molecule is being refilled again from the backscattered modes:

$$\tau_{\mathcal{D}} \gamma_{\ell,r} \gg 1. \tag{2.11}$$

There is a very elegant and powerful procedure that eliminates spurious cavity modes so that the condition (2.11) is always satisfied: one introduces *absorbing boundary conditions* (abc) in some regions of the cavity. These "surface" regions should swallow incoming signals, i.e., wave packet amplitude, and thus properly mimic escape to infinity. If the wave packets, bouncing hence and forth inside the cavity, are completely eliminated before the return time τ_D has passed, no trace of the finite cavity/electrode size will be left and $G(t, \mathbf{x}, \mathbf{x}')$ can be reconstructed. This is exactly how adding the self energy $\Sigma_{e\mathcal{M}}$ works and nothing more than this is implied, if the contacts are sufficiently large. Therefore, we are entirely free to replace the exact boundary conditions $\Sigma_{e\mathcal{M}}$ by any other ones, provided they absorb sufficiently fast (and do not disturb the immediate vicinity of the molecule-electrode junction).

This is the idea underlying the step proposed in (2.5). From what has been said above, it should have become clear that this ansatz is actually not just a good approximation, but it will give exact results if the number of metal atoms $N_{\mathcal{E}}$ is sufficiently large and the damping function has been suitably chosen. The examples given in Sect. 2.4 suggest that a relatively small number of contact atoms ($\sim$10–20) can already give reasonable results.

2.3 Toy Models

In this section we are going to analyze two toy problems as test cases, namely the conductance of an L-site tight binding wire, clean and in the presence of an obstacle that mimics a molecule.

We will begin with a single-channel wire and show that the technique introduced with (2.7) delivers the correct answer. This test case is interesting, because (a) the numerical results can be compared to analytical formulas and (b) it is particularly difficult, in the sense that the dwell time, $\tau_D \approx L/2v_F$, is (untypically) small (v_F: Fermi velocity). To illustrate the method further, we also apply it to a wire with four channels and show that not only the conductance but also more complicated quantities, like the local current density, can be obtained.

2.3.1 Single Channel Tight Binding Wire

2.3.1.1 Models and Analytical Results

The model Hamiltonian of the clean tight binding wire is given by

$$H = -t/2 \sum_{i=1}^{L} c_i^\dagger c_{i+1} + c_{i+1}^\dagger c_i \tag{2.12}$$

for spinless, non-interacting electrons. The corresponding dispersion relation reads

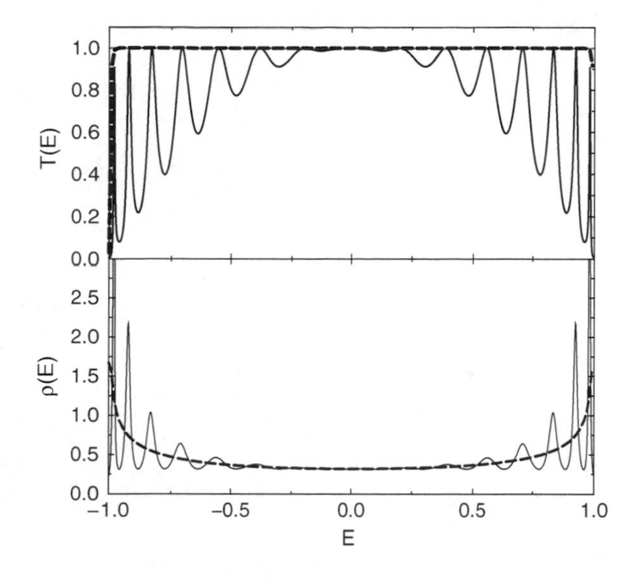

Fig. 2.6 Clean single-channel tight binding wire, *upper panel*: transmission. Analytical result (2.15) (*solid, thin*) and numerical result for $L = 256$ with fully absorbing boundary conditions (*dashed*; $i_S = 32$, $\beta = 0.3$, $\eta = 1$). Also shown is numerical result with partially absorbing boundary conditions for $L = 16$ (*solid, wiggly line*, $i_S = 1$, $\beta = \infty$, $\eta = 1$). *Lower panel*: density of states corresponding to the transmission curves shown in *upper panel*

$$\epsilon_k = -t\cos(ka), \tag{2.13}$$

where a denotes the lattice constant, and for the density of states one has

$$\varrho(E) = \frac{1}{\pi t\sqrt{1-(E/t)^2}}. \tag{2.14}$$

It exhibits the usual van Hove singularities at the edges of the band, which can also be seen in Fig. 2.6.

In order to calculate the transport coefficient of the L-chain defined in (2.12) we should couple it adiabatically to a left and a right hand reservoir. This can be done by attaching further half-infinite tight binding chains to the right and the left of the L-chain. The combined system is a perfect $1d$ crystal. Its Bloch waves travel without any backscattering through the L-chain and therefore it is a perfect conductor with transmission unity:

$$T(E) = 1 \quad |E| < t. \tag{2.15}$$

Next, let us insert an obstacle, e.g., a strong triple barrier, into the wire, a situation that still can be understood in all detail. The corresponding Hamiltonian $H_{e\mathcal{M}}$ is realized with hopping amplitudes occurring in (2.12) that take the values $t = 0.05$ at the pairs of sites $(i\pm = L/2,\ i\pm = L/2 \pm 2)$ and $t = 1$ everywhere else.

The triple barrier has two eigenstates, a symmetric and an anti-symmetric one, which are energetically nearly degenerate since the center barrier is high. The energy is given approximately by the width of the double well inside the outer barriers, $3a$. It corresponds to a wavenumber $k = \pi/3a$ which in turn implies a resonance energy close to $t/2$. Therefore, the transmission characteristics of the triple barrier should exhibit a superposition of two Lorentzians, one slightly below and one slightly above $E = t/2$. These are the features that can indeed be seen in the analytical result for the conductance (valid in the limit of weak coupling, $V \ll t$)

$$T(E) = \sum_{\alpha=\pm} \frac{1}{1+\gamma^{-2}(E-E_\alpha)^2} \tag{2.16}$$

where $E_\pm = \cos(\pi/3 \pm V \sin(\pi/3)/3)$, $t = 1$ and $\gamma = V^2 \sin(\pi/3)^3/24$. A derivation can be found in Appendix A.1. Here, we have reproduced for better clarity only an expansion of the exact answer, given in the Appendix A.1, Eq. 2.32, which is valid in the vicinity of the resonances. We also display the exact transmission in Fig. 2.7, upper panel.

2.3.1.2 Green's Function Method with Absorbing Boundary Conditions

The transmission of the combined system—wire plus triple barrier—has been given already in (2.1). In the present case it reads

$$g(E) = \mathrm{Tr}_{e\mathcal{M}}\left(G\Gamma_{\mathcal{L}}G^\dagger\Gamma_{\mathcal{R}}\right), \tag{2.17}$$

where the definition of G, (2.4), implies:

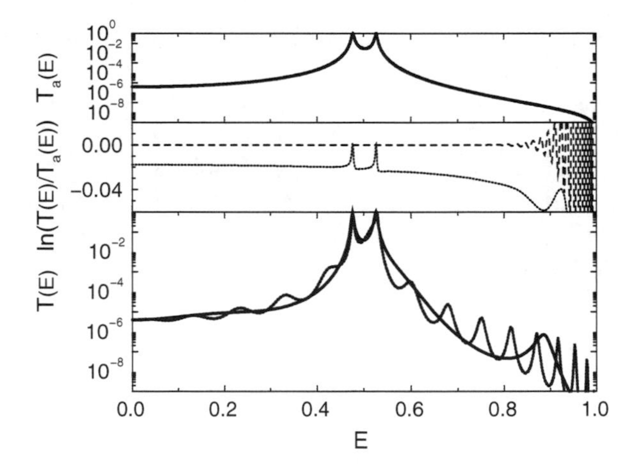

Fig. 2.7 Transmission of a single channel wire with a strong triple barrier (see text). *Upper panel*: exact analytical result. Center: deviation of numerical results from analytical calculation: $\ln\frac{T(E)}{T_a(E)}$ ($L = 256$, $i_S = 32$, $\beta = 0.3$, $\eta = 1$, *dashed*; $L = 64$, $i_S = 16$, $\beta = 0.3$, $\eta = 1$, *dotted*) with absorbing boundary conditions. *Lower panel*: partially absorbing boundary conditions (*dotted, wiggly line*: $L = 16$, $i_S = 1$, $\beta = \infty$, $\eta = 1$) (*solid, smooth line*: $L = 64$, $i_S = 1$, $\beta = \infty$, $\eta = 1$)

$$G^{-1} = E - H_{e\mathcal{M}} - \Sigma_{\mathcal{L}} - \Sigma_{\mathcal{R}}. \tag{2.18}$$

The Hamiltonian $H_{e\mathcal{M}}$ of the extended molecule is given with (2.12) and the trace $\text{Tr}_{e\mathcal{M}}$ is over the corresponding Hilbert space. The operators $\Gamma_{\mathcal{L},\mathcal{R}}$ are related to those pieces, $\Sigma_{\mathcal{L},\mathcal{R}}$, of the self energy, $\Sigma_{e\mathcal{M}}$, which describe the level broadening due to the coupling to the left ($\mathcal{L}$) and right ($\mathcal{R}$) leads:

$$\Gamma_{\mathcal{L},\mathcal{R}} = i(\Sigma_{\mathcal{L},\mathcal{R}} - \Sigma^{\dagger}_{\mathcal{L},\mathcal{R}}). \tag{2.19}$$

The precise form of $\Sigma_{\mathcal{L},\mathcal{R}}$ depends on how we couple the wire to the external leads. In the spirit of Sect. 2.3, we simply define $\Sigma_{\mathcal{L},\mathcal{R}}$ as follows:

$$\Sigma_{ij;\mathcal{L}} = i\eta_i\delta_{ij}; \quad \eta_i = \eta/(1 + \exp\beta(i - i_S)) \tag{2.20}$$

$$\Sigma_{ij;\mathcal{R}} = i\eta_i\delta_{ij}; \quad \eta_i = \eta/(1+\exp\beta(L - i - i_S)). \tag{2.21}$$

Three parameters have been introduced: η is the atomic leakage rate for all those atoms that are fully coupled to the outside; i_S describes the number of surface atoms on either side of the molecule; β is the adiabaticity parameter that models a smooth transition into the external wire.

Fig. 2.6, upper panel, displays the result of this procedure for the transmission of the clean wire. As expected, the exact result (2.15) is recovered in the case with perfectly absorbing boundary conditions (abc). For comparison, we also show a trace corresponding to incomplete absorption. The cavity modes manifest themselves in the transmission characteristics as relatively sharp resonances. In order to highlight this aspect, the lower panel of Fig. 2.6 also shows the density of states of the wire.

In Fig. 2.7 we present the transmission of the single channel wire with a triple barrier implantation. As can be seen from the upper panel, the agreement between the wire with fully abc and the analytical result is perfect. Once more we also display traces that result from a calculation with imperfectly absorbing boundaries. Traces for two different cavity sizes, $L = 16, 64$, are given. As in the previous case, Fig. 2.6, the cavity eigenmodes give rise to system size dependencies, which are the spurious resonances in the transmission characteristics. By contrast, no remnant of the system size is left if perfect abc are used, see upper panel traces for $L = 64, 256$.

Let us emphasize that the good quality of the results Figs. 2.6 and 2.7 is not a consequence of fine-tuning parameters. We have ascertained that the traces corresponding to perfectly abc are stable against variations at least in the parameter range $L = 64$–512, $\eta = 0.1$–10, $\beta = 0.03$–0.5, $i_S = 8$–64.

In the test cases presented in this subsection, analytical results have been available in order to demonstrate that the choice of parameters associated with the absorbing boundary conditions was appropriate. In more realistic situations analytical results are almost never available. Therefore, additional criteria have to be given to establish that a certain choice of boundary conditions indeed provides sufficient absorption. The basic rule is that a good implementation will yield transmission curves (largely) independent of the size $N_{e\mathcal{M}}$ of the cavity, of the choice of surface atoms inside the cavity and of the atomic leakage rate η. A calculation that satisfies these requirements, is (usually) quite reliable.

2.3.2 *Local Currents in a Many Channel Tight Binding Wire*

As a further application of our method, we calculate the local current density, j_μ, in a multi-channel wire. We begin by deriving a general formula relating j_μ to the Green's functions and self energies calculated in the preceding section. Thereafter, we shall illustrate the result by calculating the local current distribution within a double well embedded in a four channel wire.

2.3.2.1 Lattice Current Density in Terms of Green's Functions

To start with, we consider the general model Hamiltonian

$$H = \frac{1}{2}\sum_{\nu,\nu'} t_{\nu,\nu'}\, c_\nu^\dagger c_{\nu'}. \tag{2.22}$$

The multi-index ν comprises the longitudinal, i_ν, and transverse, ℓ_ν, wire coordinates. An expression for the local (longitudinal) current density may be obtained from the time dependent local density:

$$\dot{n}_\mu = \frac{i}{\hbar}[H, c_\mu^\dagger c_\mu] = -\frac{i}{2\hbar}\sum_\nu t_{\mu\nu} c_\mu^\dagger c_\nu - t_{\mu\nu}^* c_\nu^\dagger c_\mu. \tag{2.23}$$

The component of the local particle current in the longitudinal direction (right), j_μ, is given by the difference of those hopping events that enter site (i_μ, ℓ_μ) from the left and leave there again:

$$j_\mu = -\frac{i}{2\hbar}\sum_{i_\nu < i_\mu}\sum_{\ell_\nu} t_{\mu\nu} c_\mu^\dagger c_\nu - t_{\mu\nu}^* c_\nu^\dagger c_\mu. \tag{2.24}$$

The expectation value of this operator is readily expressed in terms of the Green's function [13], $G_{\nu\mu}^<(t,t') = i\langle c_\mu^\dagger(t')c_\nu(t)\rangle$,

$$\langle j_\mu\rangle = -\frac{1}{2\hbar}\int\frac{dE}{2\pi}\sum_\nu^{\mathcal{L}_\mu} t_{\mu\nu} G_{\nu\mu}^<(E) - t_{\mu\nu}^*\, G_{\mu\nu}^<(E). \tag{2.25}$$

In order to simplify the notation, we have introduced a name for sums like the one appearing in (2.24), which are restricted to the left/right half space: $\sum^{\mathcal{L},\mathcal{R}}$.

Since we only consider non-interacting fermions, $G^<_{\nu\mu}$ takes a particularly simple form (e.g., [29])

$$G^< = iG(f_{\mathcal{L}}\Gamma_{\mathcal{L}} + f_{\mathcal{R}}\Gamma_{\mathcal{R}})G^\dagger, \tag{2.26}$$

which in the limit of small voltages V leads to the following formula for the local charge current distribution $j(i_\mu) = \sum_{\ell_\mu} j_\mu$:

$$e\langle j(i_\mu)\rangle = -\frac{ie^2}{2h}V \int dE \frac{df}{dE} \sum_{\ell_\mu} \sum_{\nu}^{\mathcal{L}_\mu} t_{\mu\nu} \left[G(\Gamma_{\mathcal{L}} - \Gamma_{\mathcal{R}})G^\dagger\right]_{\nu\mu}. \tag{2.27}$$

(When writing this expression, we have assumed for simplicity that H is time reversal invariant, so that t is real and t, G and Γ are symmetric matrices. Also, $f_{\mathcal{L},\mathcal{R}}$ denote the Fermi–Dirac distribution of quasi-particles in the left and right hand reservoirs.)

2.3.2.2 Application: Double Dot Inside a Four-Channel Wire

In order to give an example for the usefulness of (2.27), we calculate j_μ for a four-channel wire with a double-well barrier. The example illustrates that the average current in the wire is a sum of contributions. They can undergo strong spatial fluctuations which are of the order of the mean current. In particular they can be positive or negative (backflow). It is only the sum of all of them which is independent of the longitudinal spatial coordinate.

The clean four-channel wire considered here is made up from four strands, which are the single-channel wires given in (2.12). These are arranged in a 4-fold cylindrical geometry and in transverse direction only nearest neighbors are being coupled, see Fig. 2.8. The coupling between all nearest neighbor pairs has been chosen as $t = 2$, except for those nine pairs that form the double well. The latter have $t = 0$. The position of these "defects" can be given in terms of the longitudinal site index, i, and the transverse index ℓ that labels the constituting single-channel wires in a clockwise fashion: $\ell = 0, 1, 2, 3$. We have switched off three couplings at and near the center ($i_c = L/2$) of the wire, i_C, $i_C \pm 2$, with site indices $\ell = 1, 2, 3$. After the definition of the model Hamiltonian, $H_{e\mathcal{M}}$, we also give the left and right contributions to the self energy $\Sigma_{e\mathcal{M}}$,

$$\Sigma^{\ell\ell'}_{ij;\mathrm{X}} = \delta_{\ell\ell'}\Sigma_{ij;\mathrm{X}}, \quad \mathrm{X} = \mathcal{L}, \mathcal{R} \tag{2.28}$$

with a self energy per strand, $\Sigma_{ij;\mathrm{X}}$, as in (2.20, 2.21).

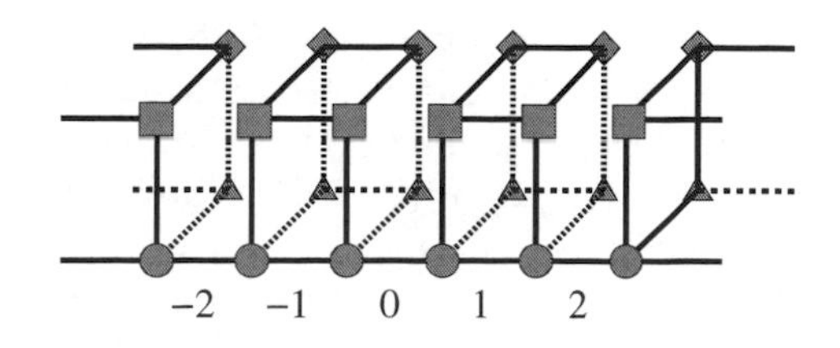

Fig. 2.8 A *double dot* inside a tight binding wire consisting of four coupled strands (indicated by different symbols)

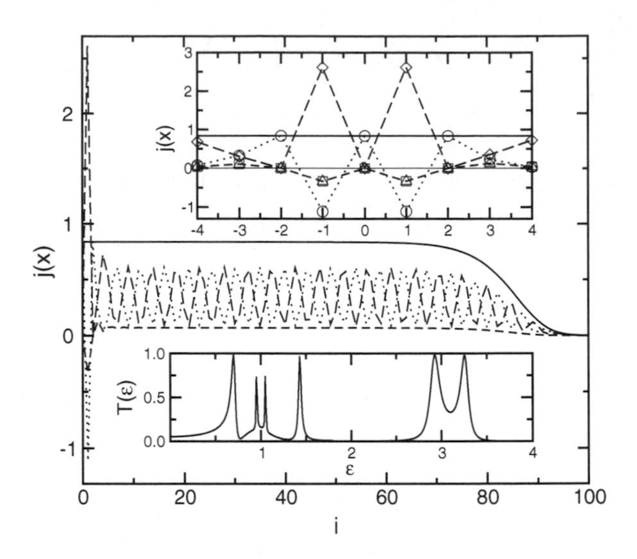

Fig. 2.9 Local current density inside a four channel wire ($L = 256$, $i_S = 32$, $\beta = 0.1$, $\eta = 1$) with a *double dot* (Fig. 2.8) at $\epsilon = 0.7$ very close to a resonance of the transmission (see lower inset). Upper inset: behavior near the *dots* (position ± 2, 0). Channels 1,3 (*open square*, *open triangle*, *dashed line*; same current density) and 2 (*open diamond*, *long dashed*) are blocked, channel zero 0 (*open circle*, *dotted line*) remains open. Inside the wells, j_2 overshoots the transmission (*solid line*) by 300%, so that local backflow in the other current channels is generated. Main: sum of local currents (*solid line*) is conserved, i.e., independent of position (except for the surface region, where by adding the self energy leakage has been introduced)

Again, we may employ (2.18, 2.19), so that the numerical evaluation of (2.27) is straightforward ($T \to 0$). Figure 2.9 shows the induced current density per applied voltage near the resonance energy, $\epsilon = 0.7$. Before we discuss this result in more detail, we first address the structure of the transmission function which is also displayed in Fig. 2.9, lower inset. It consists of three pairs of resonances, each pair resembling the symmetric and anti-symmetric eigenstate of the (isolated) double well. The pair of peaks closest to the band edges originates from hybridization of these states with those wire modes having a wavenumber matching $\pi/2_a$ and an s-wave type symmetry in transverse direction. In these two peaks the current density is homogenously distributed among the four constituting wires. The last sentence is not true for the remaining two pairs of resonances with p-wave character, where the current flows mainly in one of the wire pairs, either (0,2) or (1,3). Evidently, the two resonances closest to $\epsilon = 1$ belong to the second category, since these are much sharper and less well split than all the others.

Now, we can come back to the strong oscillations seen in the local current density of the resonance closest to the band center, cf. main panel of Fig. 2.9. There, the current flow is mostly in the (0,2) pairs. Since due to the barrier these two channels are not symmetry equivalent, the current densities j_{i0} and j_{i2} can pick up different dependencies on the longitudinal wire coordinate, i. In fact, this must be the case since at the barrier position $j_0 = 1$ while $j_2 = 0$.

The phase locking between the local currents flowing along the (0,2) pair of strands has an interesting effect on the current flow inside each well: in this region, the component j_2 acquires a value three times exceeding the average current flow. This value is compensated by a backflow in the other channels, so that a current vortex develops.

2.4 Test Cases from Quantum Chemistry Calculations

The purpose served by the tight binding calculations of the previous chapter was to demonstrate the principle. High precision in the calculations performed there was relatively easy to achieve, because the parameter space representing a perfect separation of energy (or time) scales was well accessible by numerical methods.

In quantum chemistry calculations feasible at present, the accessible system sizes often are not large enough in order to achieve such a clear scale separation. We demonstrate in this section that our method operates reasonably well, also in a practical situation, where scale separation is not perfect.

To this end, we shall consider two extreme cases, a short gold chain with a conductance $g \sim 1$ and a molecule, di-thiophenyl with $g \ll 1$. Both objects are coupled to the tip of two tetragonal bipyramids of 14 Au atoms each that represent the extension modelling a piece of the electrodes, see Fig. 2.2. We define an effective single-particle Hamiltonian, $H_{e\mathcal{M}}$, for these systems in the way explained in Sect. 2.1.

The proposed construction mechanism has been formulated in coordinate (real) space. Therefore it matches particularly well with quantum chemistry calculations using the local atomic basis sets $|X, \ell\rangle \equiv |b\rangle$ also introduced in Sect. 2.1. Adopting (2.5) to the present case, we introduce the following self energy:

$$\Sigma_{\mathrm{X}} = i\eta \sum_{b,b'}^{S_{\mathrm{X}}} |b\rangle \; [S^{-2}]_{bb'} \; \langle b'|, \quad \mathrm{X} = \mathcal{L}, \mathcal{R}. \tag{2.29}$$

The overlap matrix, $S_{bb'} = \langle b|b'\rangle$ appearing in this expression takes care of the fact that basis states belonging to different atomic sites will not be orthogonal in general. $\Sigma_{\mathcal{L},\mathcal{R}}$ is local, i.e., diagonal in the atomic basis set $|b\rangle$, in full analogy to (2.20, 2.21). Again the important input is in the strength and spatial modulation of the leakage function. In the present case, we choose it to be a constant, η, for a subset $S_{\mathcal{L},\mathcal{R}}$ of "surface atoms" and zero for all the others. In our calculations we take these sets to be the two layers of the pyramid (3×3 and 2×2) that are farthest from the molecule, see Fig. 2.2.

2.4.1 Transmission of Au Chain

We begin our analysis with the two-atom Au chain. The transmission of such chains has been studied by various groups before [12, 30]. It is well known that

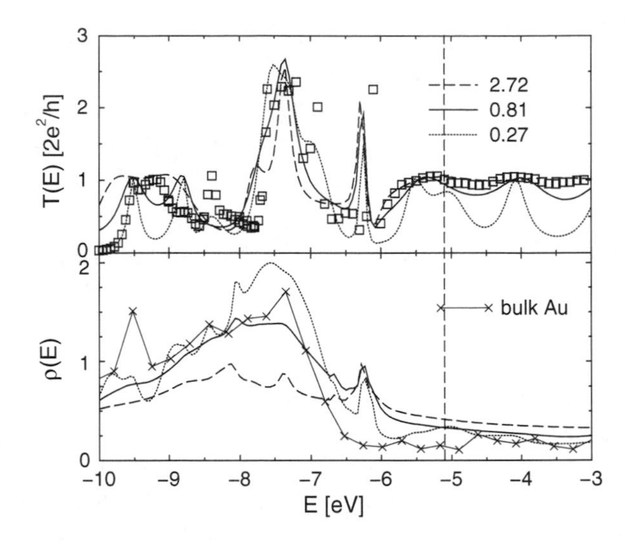

Fig. 2.10 *Upper panel*: Transmission of a two-atom Au chain linked to the tip of two tetragonal pyramids of 14 Au atoms each. Self energy (2.29) has been used. Three traces with different η are shown: 2.72 eV (*dashed*), 0.81 eV (*solid*), 0.27 eV (*dotted*). The □-symbol indicate the result with the canonical coupling and much larger pyramids with 55 atoms [3]. *Lower panel*: density of states calculated with the Green's functions used in the upper plot. Also the density of states for bulk gold is shown (symbol *times*)

reproducing the correct transmission curve is a sensitive test on the quality of the damping $\Sigma_{e\mathcal{M}}$ of the model. In Fig. 2.10, upper panel, we depict the transmission as obtained from the model (2.29). For comparison, also plotted is a result obtained from a much larger system with a self energy calculated directly from (2.3) (where $G_{\mathcal{S}}$ has been replaced by the bulk Green's function $G_{\mathcal{B}}$) [3]. Agreement for the two larger values of the coupling amongst each other and also with the original curve is established reasonably well with deviations typically less than 10%.

As was to be expected, the window of η-values in which one finds good quantitative agreement between the various curves is relatively small, simply because of the very small cavity size. In fact, the "molecule" defined by the two Au atoms in line, see Fig. 2.2, is separated from the effective surface regions $\mathcal{S}_{\mathrm{L,R}}$ by only one gold atom.

In the surface regions, the model self energy somewhat modifies the local material parameters, like the density of states (DoS). This can be seen in Fig. 2.10, lower panel. The total DoS is strongly dominated by surface atoms. It has a dependency on η that is much stronger than the one of the transmission, cf. Fig. 2.10, upper panel. Note that this modification will have a substantial impact if one were to set up a self consistent calculation with a local density obtained from the dressed Green's function (2.4). The present setup is not (and in fact does not need to be) self consistent in this sense, and therefore the modification of the surface spectral function will be without consequences for the transport calculations proposed here.

We comment on the absence of a adiabaticity parameter β in the definition of the self energy (2.29). The pyramids simulating the electrodes act as resonating cavities in the same way that the single-channel tight binding wire does, cf. Sect. 3.1. However, the tight binding wire was special in the sense that the number of surface atoms coupling to the infinite tight binding chain was only two, independent of the volume of the wire, L. For this reason the resonator modes had to be eliminated by introducing the adiabaticity parameter β. In higher dimensions, the ratio of contact surface to volume is much more favorable. Therefore, the surface damping of the resonator modes is much stronger—no real need to introduce a β-parameter here.

2.4.2 Transmission of Di-Thiophenyl

Finally, we apply our construction for the self energy to the paradigm of computational molecular electronics, the di-thiophenyl system. Again, the electrodes are modelled by the same pair of Au_{14} pyramids we have used before for the two-atom Au chain. Accordingly, the construction of the model Hamiltonian and, in particular, the self energy are just as in the previous section.

We will investigate two slightly different situations, where the sulphur atom, that ties the benzene to the Au-contact surface connects either to a single Au atom or to three of them.

2.4.2.1 S$-$Au$_1$ Coupling

The atomistic setup we consider in this subsection, is presented in Fig. 2.2. The S atom acts as the barrier that disconnects the conjugated π-system of the phenyl ring from the Au atom forming the tip of the pyramid. This atom provides the separation for the junction from the contact region, which is necessary in order to find results for the transmission independent of the choice of η within a large parameter window.

Indeed, our expectations are well confirmed by the numerical data. Fig. 2.11, upper channel, shows transmission lines for η varying over two orders of magnitude. All traces faithfully reproduce the salient resonance structures in the vicinity of 1 eV about the Fermi energy, $E_F = -5.05$ eV. In this regime, the transmission is almost unaffected by the change in η even though the average DoS changes by a factor of three, see Fig. 2.11, lower panel. Eventually, there is an impact of η in the tails of the resonances, where the transmission is small and the DoS changes with η by an order of magnitude. This regime can be better controlled by creating an additional spacer layer of Au atoms between the contact atoms and the junction.

We believe that the construction principle (2.29) is well suited to model the line broadening of resonances. There is, however, no prediction for the line shift, of

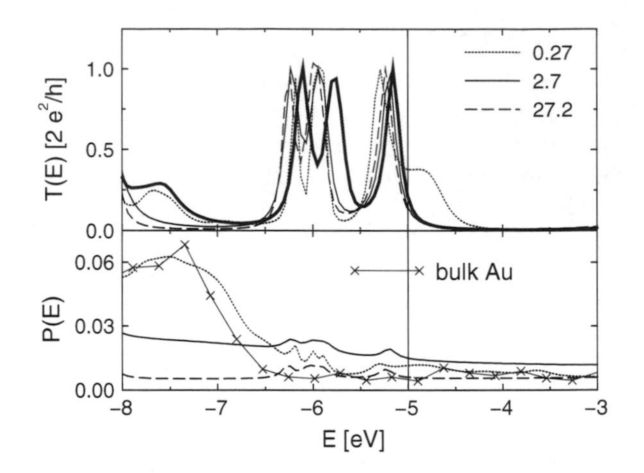

Fig. 2.11 *Upper panel*: transmission of di-thiophenyl with a S–Au_1 coupling (see Fig. 2.4). for different damping parameters $\eta = 0.27$ eV (*dotted*), 2.7 eV (*solid*) and 27.2 eV (*dashed*). Also shown 30-atom Au pyramids with conventional coupling (2.29). *Lower panel*: probability density $P(\epsilon)$ for eigenvalues (proportional to the DoS). Three traces correspond to the transmission lines of *upper panel*. Symbols $\times$ indicate $P(\epsilon)$ of bulk gold

course. Line shifts occur, e.g., when charge reorganization takes place, involving charge flow from one subsystem to another. Such processes can be reliably modelled only by increasing the number of electrode atoms in the calculation and monitoring the results.

2.4.2.2 S–Au_3 Coupling

In this paragraph, we investigate a situation in which the S atoms couple to three Au atoms rather than just one. There are two motivations for doing so. First, a two or three site configuration of the sulphur is energetically more favorable than the single site coordination used in Sect. 4.2 and thus more likely to be relevant for the understanding of experiments [3]. Second, we would like to present a simple example for a situation where the dwell time in the cavity is not sufficiently long so that the parameter window for the transmission traces being independent of the damping η has closed.

In general, one avoids coupling the S atoms directly to the surface layers, $S_{\mathrm{L,R}}$, because the change in the local DoS near the contact may feed back into the transmission. Therefore, for the S–Au_3 coupling (Fig. 2.2) we include all Au atoms into $S_{L,R}$ except for those three Au atoms that bind to S. After this minor modification, calculations proceed in the same way as before.

The resulting transmission is displayed in Fig. 2.12. For the 14-atom Au pyramids traces for three different damping values η are shown. They differ appreciably from each other and the correct result is recovered only in relatively rough terms. The limited quality of the model coupling in the present case is

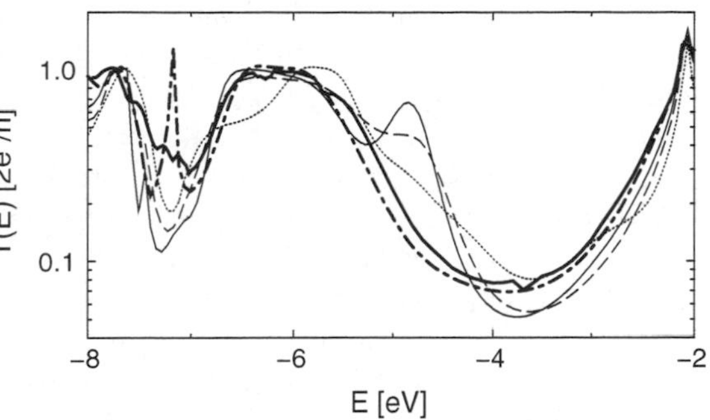

Fig. 2.12 Transmission of di-thiophenyl with a S–Au_3 coupling (see Fig. 2.4). Thin traces: 14-atom Au pyramids with $\eta = 0.81$ eV (*dotted*), 2.76 eV (*solid*) and 8.1 eV (*dashed*). Fat traces: 55-atom Au pyramids with $\eta = 2.7$ eV (*dot-dashed*) and conventional coupling (2.29) (*solid*)

expected. Since the number of surface atoms that couples to the S–Au_3 unit (eight) is much bigger than with the S–Au_1 coupling (four), the dwell time is reduced too strongly for our simple model (2.5) to work with high precision.

The situation is much more favorable if larger cavities are used. Fig. 2.12 also shows the result for a 55-atom Au cluster, which is in good agreement with the conventional calculation.

2.5 Discussion and Outlook

The method relying on absorbing boundary conditions (abc) we have outlined in the preceding sections, has an important advantage over the more orthodox way of calculating an essentially exact self energy (2.3): it is much easier to handle. If the abc calculation meets the requirements listed in Sect. 2.1, both methods are expected to yield identical results for the Green's functions.

We have demonstrated how the abc approach can be combined with standard quantum chemistry calculations for extended molecules, $e\mathcal{M}$. Because it is not necessary any more to calculate self energies, $\Sigma_{e\mathcal{M}}$, transport calculations based on the Landauer–Büttiker theory are greatly simplified.

The computational effort is dominated by the quantum chemistry calculation for $e\mathcal{M}$. Since the number of electrode atoms, $N_{\mathcal{E}}$, that need to be included in the calculation, is similar for both methods, one expects that the computational effort is roughly the same.

The number $N_{\mathcal{E}}$ is too large in order to allow highly correlated, essentially exact methods to be used for calculating $G_{e\mathcal{M}}$. However, calculations based on density functional theory can be done very efficiently for these system sizes. Furthermore, Hartree–Fock calculations are within reach. The last point is of particular interest, because this is a prerequisite to test functionals against each other (BP89, B3Lyp, LHF, etc.), that have a very different degree of self interactions.

These tests are important for the recent debate on the origin of the discrepancy between theoretical and experimental results on the conductance observed for several organic molecules [2–6]. The conductance of mono atomic chains can be investigated very well with the combination of density functional theory (DFT) and Landauer–Büttiker formula, that have been employed in Sect. 2.4 [31]. By contrast, theoretical expectation for the transmission of organic molecules tend to deviate by one or more orders of magnitude from the experimental findings [3].

Apart from experimental difficulties, approximations implicit to the DFT based transport formalism, could very well be responsible for this discrepancy, because the Green's functions that derive from the (ground state) Kohn–Sham-formalism do not necessarily provide a good description of the system dynamics. This description can indeed be acceptable, if electron density is relatively smooth and the system is at least close to metallic, as it is in single-atom metal chains. Under these conditions, single-particle wavefunctions, which are essentially plane wave states, give a good representation of the spatial properties of the true Green's function. Then the (local) spectral properties (i.e., the band structure of the atomic wire) primarily determine the transport characteristics and those can be given quite accurately by present days DFT calculations.

However, organic molecules are a different case, because at least in the vicinity of the contact region, the electron density is far from homogenous. In these regions, the spatial structure of the Kohn–Sham Green's function will in general not be a faithful representation of reality, certainly not within local or semi-local approximations of the exchange-correlation functional. Since these very regions form exactly the bottleneck for the transport current, theoretical calculations employing such functionals cannot necessarily be expected to be very precise. One can hope that the usage of non-local functionals will improve this situation. Which of the various non-local terms that exist in the exact (quasi-static non-equilibrium) functional, is the most important one and how to implement it in practical transport calculations, are at present two of the most thrilling issues in the field of molecular electronics.

Note Added in Proof Several aspects of the analysis in Sect. 2.3 are nicely corroborated by a recent study on modeling extended contacts for carbon devices [32].

Acknowledgments The authors thank K. Busch for drawing their attention to current applications of absorbing boundary conditions in the field of quantum optics. Also, they are indebted to F. Weigend for his help in using the program package TURBOMOLE and numerous instructive discussions. Finally, they express their gratitude to P. Wölfle for helpful comments on the manuscript and in particular for his continuous support of this work. Support has also been received from the *Center for Functional Nanostructures* at Karlsruhe Institute of Technology (KIT) and is gratefully acknowledged.

A.1 A Triple Barrier in a Tight Binding Wire

A one-dimensional tight binding chain with only nearest neighbor hopping and three barriers is considered.

$$\times - \times - \times - \times + \times - \times - \times + \times - \times - \times + \times - \times - \times - \times$$

... M−2 M−1 M V_L M+1 ... N V_C N+1 ... O V_R O+1 O+2 O+3 ...

The Hamiltonian of this problem is:

$$H_{i,j} = t_i\left(\delta_{i,j+1} + \delta_{i+1,j}\right) \tag{2.30}$$

with hopping amplitudes t_i given by

$$\begin{array}{ll} t_M = V_L & \text{left barrier} \\ t_N = V_C & \text{center barrier} \\ t_O = V_R & \text{right barrier} \\ t_i \ = 1 & \text{everywhere else} \end{array}$$

and barriers located at positions M, N, O. The scattering states in the four different sections of the wire can be written as

$$\begin{array}{ll} \Psi_j = e^{ikj} + re^{-ikj} & \text{left lead} \\ \Psi_j = ae^{ikj} + be^{-ikj} & \text{left to center barrier} \\ \Psi_j = ce^{ikj} + de^{-ikj} & \text{center to right barrier} \\ \Psi_j = te^{ikj} & \text{right lead} \end{array}$$

with energy $E(k) = -2\cos k$ (length scales are measured with respect to a). Let us now write down the consequences of the Schrödinger equation left and right of each barrier:

$$\begin{aligned}
E\Psi_M &= \Psi_{M-1} + V_L\Psi_{M+1} \\
E\left(e^{Mik} + re^{-Mik}\right) &= e^{(M-1)ik} + re^{(1-m)ik} + V_L\left(ae^{(M+1)ik} + be^{-(M+1)ik}\right) \\
E\Psi_{M+1} &= V_L\Psi_M + \Psi_{M+2} \\
E\left(ae^{(M+1)ik} + be^{-(M+1)ik}\right) &= V_L\left(e^{Mik} + re^{-Mik}\right) + ae^{(M+2)ik} + be^{-(m+2)ik} \\
E\Psi_N &= \Psi_{N-1} + V_C\Psi_{N+1} \\
E\left(ae^{Nik} + be^{-Nik}\right) &= ae^{(N-1)ik} + be^{(1-N)ik} + V_C\left(ce^{(N+1)ik} + de^{-(N+1)ik}\right) \\
E\Psi_{N+1} &= V_C\Psi_N + \Psi_{N+2} \\
E\left(ce^{(N+1)ik} + de^{-(N+1)ik}\right) &= V_C\left(ae^{Nik} + be^{-Nik}\right) + ce^{(N+2)ik} + de^{-(N+2)ik} \\
E\Psi_O &= \Psi_{O-1} + V_R\Psi_{O+1} \\
E\left(ce^{Oik} + de^{-Oik}\right) &= ce^{(O-1)ik} + de^{(1-O)ik} + V_R te^{(O+1)ik} \\
E\Psi_{O+1} &= V_R\Psi_O + \Psi_{O+2} \\
Ete^{(O+1)ik} &= V_R\left(ce^{Oik} + de^{-Oik}\right) + te^{(O+2)ik}
\end{aligned}$$

or in matrix notation:

$$\begin{pmatrix} e^{(M+1)ik} & e^{-(M+1)ik} \\ V_L e^{Mik} & V_L e^{-Mik} \end{pmatrix} \begin{pmatrix} 1 \\ r \end{pmatrix} = \begin{pmatrix} V_L e^{(M+1)ik} & V_L e^{-(M+1)ik} \\ e^{Mik} & e^{-Mik} \end{pmatrix} \begin{pmatrix} a \\ b \end{pmatrix}$$

$$\begin{pmatrix} e^{(N+1)ik} & e^{-(N+1)ik} \\ V_C e^{Nik} & V_C e^{-Nik} \end{pmatrix} \begin{pmatrix} a \\ b \end{pmatrix} = \begin{pmatrix} V_C e^{(N+1)ik} & V_C e^{-(N+1)ik} \\ e^{Nik} & e^{-Nik} \end{pmatrix} \begin{pmatrix} c \\ d \end{pmatrix}$$

$$\begin{pmatrix} e^{(O+1)ik} & e^{-(O+1)ik} \\ V_R e^{Oik} & V_R e^{-Oik} \end{pmatrix} \begin{pmatrix} c \\ d \end{pmatrix} = t e^{(O+1)ik} \begin{pmatrix} V_R \\ E - e^{ik}. \end{pmatrix}$$

Since our interest is in the transmission, $T = |t|^2$, we only extract an equation for the transmission coefficient, t:

$$\begin{aligned} &\Big[(V_L{}^2 e^{ik} - e^{-ik})(V_C{}^2 e^{ik} - e^{-ik}) - (V_L{}^2 e^{-(2M+1)ik} - e^{-(2M+1)ik}) \\ &(V_C{}^2 e^{(2N+1)ik} - e^{(2N+1)ik})\Big](V_R{}^2 e^{ik} - e^{-ik}) \\ &- \Big[(V_L{}^2 e^{ik} - e^{-ik})(V_C{}^2 e^{-(2N+1)ik} - e^{-(2N+1)ik}) \\ &-(V_L{}^2 e^{-(2M+1)ik} - e^{-(2M+1)ik})(V_C{}^2 e^{-ik} - e^{ik})\Big](V_R{}^2 e^{(2O+1)ik} - e^{(2O+1)ik}) \\ &= \frac{8}{t} V_L V_C V_R \sin^3 k. \end{aligned} \tag{2.31}$$

After specializing to the case of a strong, symmetric barrier, where $V_L = V_C = V_R = V \ll 1$ and $M = 0$, $N = 2$, $O = 4$, we conclude

$$T(k) = \left(2V^3 \frac{\sin^3(ka)}{|\sin^2(3ka) - V^2 \sin^2(2ka) + 2V^2 e^{ika} \sin(3ka)\sin(2ka) + O(V^4)|}\right)^2 \tag{2.32}$$

which is also displayed in Fig. 2.7, upper panel, in the main body of the paper, where also a brief discussion of this result may be found.

References

1. Likharev, K., Mayr, A., Mickra, I., Türel, Ö.: Ann. N. Y. Acad. Sci. **1006**, 146–163 (2003)
2. Reimers, J.R., Cai, Z.-L., Bilić, A., Hush, N.S.: Ann. N. Y. Acad. Sci. **1006**, 235–251 (2003)
3. Evers, F., Weigend, F., Koentopp, M.: Phys. Rev. B **69**, 235411 (2004)
4. Kurth, S., Stefanucci, G., Almbladh, C.-O., Rubio, A., Gross, E.K.U.: Phys. Rev. B **72**, 035308 (2005)
5. Sai, N., Zwolak, M., Vignale, G., Di Ventra, M.: Phys. Rev. Lett. **94**, 186810 (2005)
6. Koentopp, M., Burke, K., Evers, F.: Phys. Rev. B **73**, 121403(R) (2006)
7. Homer: Odyssee, 12. Gesang, Vers, pp. 201–259
8. Bohr, D., Schmitteckert, P., Woelfle, P.: Europhys. Lett. **73**(2), 246–252 (2006)
9. Landauer, R.: IBM J. Res. Dev. **1**, 233 (1957)
10. Büttiker, M., Imry, Y., Landauer, R., Pinhas, S.: Phys. Rev. B **31**, 6207 (1985)

11. Xue, Y., Datta, S., Ratner, M.A.: J. Chem. Phys. **115**, 4292 (2001)
12. Brandbyge, M., Mozos, J.-L., Ordejon, P., Taylor, J., Stokbro, K.: Phys. Rev. B **65**, 165401 (2002)
13. Meir, Y., Wingreen, N.S.: Phys. Rev. Lett. **68**, 2512 (1992)
14. Fisher, D.S., Lee, P.A.: Phys. Rev. B **23**, 6851 (1981)
15. Baranger, H.U., Stone, A.D.: Phys. Rev. B **40**, 8169–8193 (1989)
16. Ahlrichs, R., Bär, M., Häser, M., Horn, H., Kölmel, C.: Chem. Phys. Lett. **162**, 165 (1989)
17. Eichkorn, K., Treutler, O., Oehm, H., Haeser, M., Ahlrichs, R.: Chem. Phys. Lett. **242**, 652 (1995)
18. Eichkorn, K., Weigend, F., Treutler, O., Ahlrichs, R.: Theor. Chem. Acc. **97**, 119 (1997)
19. Park, J., Pasupathy, A.N., Goldsmith, J.I., Chang, C., Yaish, Y., Petta, J.R., Rinkoski, M., Sethna, J.P., Abruña, H.D., McEuen, P.L., Ralph, D.C.: Nature **417**, 722 (2002)
20. Liang, W., Shores, M.P., Bockrath, M., Long, J.R., Park, H.: Nature **417**, 725 (2002)
21. Heersche, H.B., De Groot, Z., Folk, J.A., Kouwenhoven, L.P., van der Zant, H.S.J., Houck, A.A., Labaziewicz, J., Chuang, I.L.: Phys. Rev. Lett. **96**, 017205 (2006)
22. Abrikosov, A.A., Gorkov, L.P., Dzyaloshinkski, I.E.: Methods of Quantum Field Theory in Statistical Physics. Dover, New York (1963)
23. Mahan, G.D.: Many Particle Physics, 3rd edn. Kluwer/Plenum Publishers (2000)
24. Brenig, W.: Statistical Theory of Heat: Nonequilibrium Phenomena, vol. II. Springer, Berlin (1989)
25. Schollwöck, U.: Rev. Mod. Phys. **77**, 259–315 (2005)
26. Perdew, J.P.: Phys. Rev. B **33**, 8822 (1986)
27. Becke, A.D.: Phys. Rev. A **38**, 3098 (1988)
28. Schäfer, A., Huber, C., Ahlrichs, R.: J. Chem. Phys. **100**, 5829 (1994)
29. Datta, S.: Electronic Transport in Mesoscopic Systems. Cambridge University Press, Cambridge (1995)
30. Palacios, J.J., P'erez-Jim'enez, A.J., Louis, E., SanFabi'an, E., Verg'es, J.A.: Phys. Rev. B **66**, 035322 (2002)
31. Agrait, N., Yeyati, A.L., van Ruitenbeek, J.M.: Phys. Rep. **377**, 81–279 (2003)
32. Nemec, N., Tomanek, D., Cuniberti, G.: Phys. Rev. B **77**, 125420 (2008)

Chapter 3
Recent Advances in Studies of Current Noise

Yaroslav M. Blanter

3.1 Introduction

Current noise in the last decade proved to be an efficient means of investigation of nanostructures. Currently, it is a broad field, with over a hundred groups, experimental as well as theoretical, and is actively developing. This short review presents a brief introduction to the field, concentrating on recent developments. There is a large amount of literature available. General introduction to noise in solids can be found in the broad scope book by Kogan [1]. An extensive review on shot noise was written by Büttiker and the author [2]. A collection of shorter review articles intended to summarize the main directions of the field was published as proceedings of the NATO ARW held in Delft in 2002 [3]. These publications cover the field comprehensively, and there is no need to repeat all the material. This article is intended for researchers wishing to enter the field. We will only give a brief introduction to the subject of shot noise, turning then to recent developments of the field. The size of this article makes it impossible to describe all research. We will therefore self-impose the following limitations. First, we only consider papers published in 2000 and later—everything before that date can be found in Ref. [2]. Second, the choice of topics is mainly related to the experimental breakthroughs. Many papers of excellent quality will have to stay outside the framework of this article. Additionally, on purpose we do not discuss here very important issues of entanglement and the theory of measurement: In our opinion, they are better discussed in connection with properties of qubits, and we do not have enough space here for a comprehensive review of the field.

Y. M. Blanter (✉)
Kavli Institute of Nanoscience, Delft University of Technology, Lorentzweg 1,
2628 CJ Delft, The Netherlands
e-mail: y.m.blanter@tudelft.nl

M. Vojta et al. (eds.), *CFN Lectures on Functional Nanostructures – Volume 2*,
Lecture Notes in Physics 820, DOI: 10.1007/978-3-642-14376-2_3,

Now, we give a very brief overview of results well established in the field. Current through any nanostructure fluctuates in time. There are at least two reasons for these fluctuations: (i) thermal fluctuations of occupation numbers in the reservoirs; (ii) randomness of transmission and reflection of electrons. At equilibrium, only the former are important, and one has *Nyquist–Johnson noise*. We define the noise spectral power,

$$S(\omega) = \left\langle \delta\hat{I}(t)\delta\hat{I}(t') + \delta\hat{I}(t')\delta\hat{I}(t) \right\rangle_{\omega}, \tag{3.1}$$

where $\delta\hat{I}(t) \equiv \hat{I}(t) - \langle I \rangle$, $\hat{I}$ is the current operator, and the averaging is both quantum-mechanical and statistical over the states in the reservoirs. Nyquist noise $S(0) = 4Gk_BT$, with G being the conductance of a nanostructure, just follows from the fluctuation–dissipation theorem.

At zero temperature, only fluctuations due to the randomness of scattering are important. They are known as *shot noise* and can be expressed [4–7] in terms of the transmission eigenvalues $\{T_p\}$ of the nanostructure, where p labels the transport channels,

$$S(0) = \frac{2_s e^3 |V|}{\pi\hbar} \sum_p T_p(1 - T_p), \tag{3.2}$$

and 2_s is the number of spin projections. Fully open ($T_p = 1$) and fully closed ($T_p = 0$) channels do not produce any noise, since scattering is not random: electrons are either fully reflected or fully transmitted.

To appreciate (3.2), we need a reference point. The latter was provided as early as 1918 by Schottky. Consider the *Poisson process*: transmitted electrons leave the barrier in a random and uncorrelated fashion. In other words, the current is expressed as $I(t) = e \sum \delta(t - t_n)$, where t_n are random uncorrelated quantities, with the average interval τ between arrivals of consecutive electrons. The average current is $I = e/\tau$, while the current noise for this process is $S(\omega) = 2eI$ and does not depend on frequency. This *Poisson value* $S_P = 2eI$ gives us the reference point. Taking into account the Landauer formula for conductance,

$$G = \frac{2_s e^2}{2\pi\hbar} \sum_p T_p,$$

we see that the actual noise (3.2) is always suppressed with respect to S_P. This suppression is characterized by the *Fano factor* $F \equiv S(0)/S_P$, which can vary between zero and one.

Note that Nyquist and shot noises are in fact the two limiting cases of the same phenomenon. One can express this noise in terms of the transmission eigenvalues. There are other types of noise, which are always or often present in nanostructures, and which are not related to transmission properties. The most common example is low-frequency noise, proportional to the square of the applied voltage and inversely proportional to the frequency. The origin of this noise is not universal and usually is attributed to slow motion of impurities in the substrate. Such transport-unrelated noises are not considered here.

Let us now mention the basic properties of shot noise. More details can be found in Ref. [2] and references cited therein.

- For basic types of nanostructures, the Fano factor assumes universal values: $F = 1$ for a tunnel barrier, $F = 1/2$ for a symmetric double barrier, $F = 1/3$ for a diffusive wire, $F = 1/4$ for a symmetric chaotic cavity, $F = 0$ for a ballistic conductor (for instance, a quantum point contact in the plateau regime). These results have been derived theoretically by various means and confirmed experimentally.
- These results are classical; quantum mechanics only enters for calculation of transmission eigenvalues and in quantum (Fermi) statistics of electrons. For this reason, many results can be reproduced by purely classical methods, based on Boltzmann or rate equations with Langevin random forces.
- Notion of noise can be generalized to multi-terminal conductors. Current correlations calculated at different terminals are always negative. This follows from the fact that electrons obey Fermi statistics.
- The Fano factor is proportional to the electron charge. This concept can be generalized to the situation when current is carried by fermionic quasiparticles. For instance, transport between a normal metal and a superconductor for voltages below the superconducting gap is only possible by means of Andreev reflection, and is associated with the charge transfer in quanta of $2e$. This gives $F = 2$. Another example is transport in a quantum Hall bar over a barrier, which is associated with the charge e/q for the filling factor $\nu = e/q$. The Fano factor in this case becomes $F = 1/q$, which also has been measured experimentally. Generally, shot noise can be used to determine the quasiparticle charge.
- Effect of interactions of shot noise can be very different. Dephasing does not have any effect on noise, unless, of course, one discusses a phase-sensitive effect like Aharonov-Bohm oscillations. Electron–electron interactions result in heating, increasing the Fano factor, for instance, to the value $F = \sqrt{3}/4$ in diffusive wires instead of $F = 1/3$. Electron–phonon interactions suppress shot noise down to zero, since the energy is taken out of the system. This is why there is no shot noise in macroscopic systems. All these considerations assume that the ground state of a conductor is Fermi-liquid-like. If interactions lead to a formation of a new state, the situation can be very different.
- Both shot and Nyquist noises are white—frequency independent in a wide interval of frequency. For non-interacting electrons, the frequency dependence appears at the quantum scale $\hbar\omega \sim k_B T, e|V|$. For instance, at zero temperature shot noise has the following form,

$$S(\omega) = \frac{2_s e^2}{\pi\hbar} \begin{cases} \hbar|\omega| \sum_p T_p^2 + e|V| \sum_p T_p(1 - T_p), & \hbar|\omega| < e|V|; \\ \hbar|\omega| \sum_p T_p, & \hbar|\omega| > e|V|. \end{cases} \tag{3.3}$$

The part growing as $|\omega|$ is known as quantum noise. Other energy scales come from electron–electron interaction; in the most common case, the scale is just inverse *RC*-time—the time scale for classical charge relaxation.

Real life is fortunately more complicated than these simple rules, and this is why current noise is still a subject of active research. Below we consider a number of phenomena which go beyond these rules and are currently in the focus of attention. We specifically concentrate on four topics: non-symmetrized cumulants and quantum noise; counting statistics; super-Poissonian noise; and current noise and interferometry. For each of these subjects, we outline the main ideas, describe the experiments, and provide the full collection of references to the original papers.

Prior to that we would like to mention a number of experimental developments of fundamental importance over the last 5 years. They confirmed existing theoretical predictions, and generated a subsequent stream of literature, but due to the space limitations we cannot discuss them in detail.

- Observation of shot noise suppression (cold and hot electrons) in chaotic cavities [8]; crossover from classical ($F = 0$) to quantum $F = 1/4$ shot noise in chaotic cavities by tuning the dwell-time [9].
- Doubling of the Fano factor at the interface between a normal metal and a superconductor [10]; also in the presence of finite-frequency field (photon-assisted effect) [11].
- Very clean observations of giant shot noise in SNS junctions due to multiple Andreev reflection (MAR) [12]; crossover from MAR regime to noise of the quasiparticle current [13]; shot noise in the regime of coherent and incoherent MAR in disordered SNS junctions [14]; MAR in superconductor–semiconductor–superconductor junctions [15].
- Noise in an array of quantum dots [16]; noise in an array of chaotic cavities formed by point contacts [17].
- Shot noise suppression in hopping conduction [18, 19].
- Noise for photon-assisted tunneling [20].
- Noise in quantum dots in the Coulomb blockade regime [21, 22].

3.2 Quantum Noise

In (3.1), we defined noise power as a symmetric correlator. With this definition, $S(\omega)$ is always even in frequency. Indeed, a classical detector does not know anything about the order of the current operators and cannot distinguish between positive and negative frequencies: It only measures a symmetric combination. Can we measure non-symmetrized noise,

$$S_q(\omega) = 2\int dt\, e^{-i\omega(t-t')}\langle \hat{I}(t)\hat{I}(t')\rangle \ ? \tag{3.4}$$

For such measurement, we obviously need a quantum detector. Let us illustrate the basic notions with an example of a detector which is a two-level system [8, 24], with the states $|a\rangle$ (energy E_a) and $|b\rangle$ (energy E_b). Interaction between the detector and the system is supposed to be weak and proportional to the current operator, $\hat{H} = \alpha|b\rangle\langle a|\hat{I}(t) + h.c.$ The transition rates between the two states of the detector follow from the Fermi golden rule,

$$\Gamma_{a\to b} = \frac{|\alpha|^2}{2\hbar^2} S_q\left(\frac{E_b - E_a}{\hbar}\right). \tag{3.5}$$

Thus, if one measures the transition rate *at zero frequency*, the result yields asymmetric noise correlator (3.4) *at finite frequency*. For $E_b > E_a$, the detector absorbs energy from the noise; otherwise, it emits energy. Thus, noise at positive/negative frequency corresponds to absorption/emission, respectively. It is not symmetric, since transition rates are not the same for emission and absorption. For instance, at equilibrium these transition rates obey the detailed balance, $\Gamma_{a\to b}p_a = \Gamma_{b\to a}p_b$, where p_a and p_b are the occupation probabilities of the detector states, which obey the Boltzmann distribution. We obtain

$$S_q(\omega)/S_q(-\omega) = \exp(-\hbar\omega/k_BT).$$

At zero temperature, $S(\omega) = 0$ for positive frequencies: There is no energy that detector can absorb from noise. For non-equilibrium noise, at zero temperature absorption is only possible if the energy provided by the external voltage is high enough, $\hbar\omega < e|V|$. The result at zero frequency expressed in terms of conductance G and the Fano factor F of the nanostructure is

$$S_q(\omega) = 2G\begin{cases} -2\hbar|\omega, & \hbar\omega < -e|V|; \\ (e|V| - \hbar\omega) - (1-F)(e|V| + \hbar\omega), & -e|V| < \hbar\omega < 0; \\ F(e|V| - \hbar\omega), & 0 < \hbar\omega < e|V|; \\ 0, & e|V| < \hbar\omega. \end{cases} \tag{3.6}$$

First measurement of non-symmetrized noise using a quantum detector was performed by Deblock et al. [25] who used a Josephson junction (JJ) as a detector. If JJ is biased at a constant voltage, there is no dissipative (quasiparticle) current at voltages lower than $2\Delta/e$, where Δ is the superconducting gap. Such current could come from MAR, however, if the transparency of the insulating layer between two superconductors is low, MAR can be neglected. Thus, the dissipative current is zero for $eV < 2\Delta$ and grows linearly for higher voltages. However, if the junction is submitted to external radiation, the situation is different: an electron can absorb a phonon with the frequency ω. Provided the voltage is between $(2\Delta - \hbar\omega)/e$ and $2\Delta/e$, this absorption will result in the quasiparticle dc current. The amplitude of the current depends on the amplitude of the external radiation. In particular, if the external radiation originates from current noise produced by a nanostructure, the quasiparticle current depends linearly on the *non-symmetrized* spectral power of the current noise [25, 26],

$$I_{\text{pat}}(V) = \int_0^\infty \frac{d\omega}{2\pi}\left(\frac{e}{\hbar\omega}\right)^2 S_{qV}(-\omega) I_{qp}\left(V + \frac{\hbar\omega}{e}\right), \quad eV < 2\Delta,$$

where I_{qp} (V) is the quasiparticle current without external radiation, the photo-assisted current $I_{\text{pat}}(V)$ is the dc current in the presence of external noise, and S_{qV} $(-\omega)$ is the non-symmetrized voltage noise, which is related to the current noise via the impedance of the circuit. Sweeping the bias voltage V and measuring the current, one can restore the frequency dependence of the non-symmetrized correlator. A great advantage of such a detector is that the detector itself at $eV < 2\Delta$ is noiseless, since there is no quasiparticle current.

To demonstrate the possibility of detection of non-symmetric noise, Deblock et al. [25] measured noise produced by a Cooper pair box (also known as superconducting charge qubit). This is a double junction superconducting structure with two close levels corresponding to the states with N and $N + 1$ Cooper pairs in the box, respectively. All other states of the system lie far away from these two states and can be ignored. The splitting ϵ between the states N and $N + 1$ can be tuned with the gate voltage; the minimal value of this splitting is $\epsilon = E_J$, with E_J being the Josephson energy, achieved for $Q \equiv C_g V_G = e$, where C_g is the capacitance to the gate. In an ideal system, the current noise is determined from (5): only one transition is possible, with the frequency $\epsilon/\hbar$, and thus the current noise has a delta peak around this frequency. In reality, one has to take into account that the levels are broadened by tunneling, and thus the noise sharply peaks around the frequency $\omega_0 = \sqrt{\epsilon^2 + \Gamma^2}/\hbar$, with Γ being the tunnel rate [27]. One speaks of the *quasiparticle peak* in noise, with $Q < e$ and $Q > e$ corresponding to emission and absorption, respectively. This is valid in the coherent regime $E_J \gg \Gamma$; in the opposite incoherent regime, $E_J \ll \Gamma$, one has a broad peak around zero frequency. The experimental observation of Ref. [25] was that noise on the emission side of the quasiparticle peak is much stronger that noise on the absorption side, thus confirming that the quantity measured is the non-symmetrized current correlator.

A clear measurement of noise for a broad interval of frequencies which could demonstrate the crossover from zero noise at the absorption side to white shot noise at low frequencies and further to quantum noise at the emission side is still not available in the literature. However, there is further data demonstrating the detection of quantum noise. In Ref. [26], one Josephson junction was used as a noise source, and another one as a detector. The source was biased at voltages above $2\Delta/e$ and thus produced white shot noise from the quasiparticle current. The detected noise was twice as low as the full expected shot noise of the quasiparticle current, which shows that the non-symmetrized correlator was measured: only $S_q(-\omega)$, but not the contribution from $S_q(\omega)$.

Another detector used in the experiments is a quantum dot [28] at low bias. Without external noise, current only flows when an electron level lies in the window between the chemical potentials of the reservoirs. As the function of the gate voltage, current has a peak. With the external noise, tunneling via excited states becomes possible, and additional peaks in the current appear. In the experiments, the magnitude of these additional peaks at the emission side was clearly stronger than the one at the emission side.

3.3 Counting Statistics

Shot noise originates from the random nature of electron transfers. One can, at least in principle, count these transfers in real time, and from the results of the measurements deduce the average current and current noise. We have seen that shot noise contains some information about scattering properties of the nanostructure which cannot be obtained from the conductance. Higher current moments can also be deduced from the same measurement and may contain even more information. They are described with the notion of *full counting statistics* (FCS).

Let us proceed with a bit of the probability theory. Suppose we make a measurement counting some random events—for instance, electron transfers through a barrier—during a certain time interval Δt. The number of events N measured during the time interval is a *random number*, characterized by the probability P_N that precisely N events will be observed in a measurement. If one repeats identical measurements M_{tot} times and counts the number of measurements M_N that give the count N, the ratio M_N/M_{tot} gives the probability P_N in the limit $M_N \gg 1$. This probability distribution is normalized, $\sum_N P_N = 1$. Once we know it, we can estimate the average of any function f_N,

$$\langle f \rangle = \sum_N f_N P_N .$$

The description of the statistics with the distribution function P_N is not always the most convenient one. The problem is that if we measure first during the time interval Δt_1 (distribution function P_1) and then during Δt_2 (P_2), the distribution function for the total interval $\Delta t = \Delta t_1 + \Delta t_2$ is a convolution (provided the two intervals are independent),

$$P_N^{\mathrm{tot}} = \sum_{M=0}^{N} P_{1,M} P_{2,N-M} .$$

Most conveniently, this is expressed in terms of the *characteristic function* of a probability distribution,

$$\Lambda(\chi) = \left\langle e^{i\chi N} \right\rangle = \sum_N P_N e^{i\chi N} .$$

For independent events, the characteristic function of the total distribution is just a product of characteristic functions of each type of events, $\Lambda^{\mathrm{tot}}(\chi) = \Lambda_1(\chi)\Lambda_2(\chi)$. The function $\ln \Lambda(\chi)$ is thus proportional to the duration of the measurement Δt. Differentiating this function k times with respect to $i\chi$ and setting subsequently $\chi = 0$, we generate the kth cumulant of the distribution.

Thus, the first derivative produces the average N, and the second derivative reproduces the variance.

For electron transfers in nanostructures, it is customary to consider statistics of charge $Q = eN$ transmitted from the left to the right during the time interval Δt. We assume that this measurement time is long enough, so that $Q \gg e$ and the laws of statistics apply. On average, $\langle Q \rangle = \langle I \rangle \Delta t$. The second cumulant gives the shot noise at zero frequency,

$$\langle\langle Q^2 \rangle\rangle \equiv \langle Q^2 \rangle - \langle Q \rangle^2 = \frac{\Delta t S(0)}{2}.$$

The characteristic function of the transmitted charge can be expressed in terms of transmission eigenvalues of the nanostructure [29–31],

$$\ln \Lambda(\chi) = 2_s \Delta t \int \frac{dE}{2\pi\hbar} \sum_p \ln\{1 + T_p(e^{i\chi} - 1) f_L(E)[1 - f_R(E)] + T_p(e^{-i\chi} - 1) f_R(E)[1 - f_L(E)]\}. \tag{3.7}$$

The logarithm of the characteristic function is a sum over transport channels, suggesting that electron transfers in different channels and over different energy intervals are independent. Differentiating this expression once and twice over the counting field χ, we reproduce the Landauer formula and the expression for the current noise. At zero temperature, (3.7) becomes

$$\ln \Lambda(\chi) = \pm \frac{2_s eV\Delta t}{2\pi\hbar} \sum_p \ln\left[1 + T_p\left(e^{\pm i\chi} - 1\right)\right], \tag{3.8}$$

where the upper and lower signs refer to the case of positive and negative voltages, respectively. Let us for simplicity consider $V > 0$. We define the number of attempts $N_{\text{at}} = 2_s \Delta t eV/(2\pi\hbar)$ and assume it to be integer. Equation 3.8 for one transport channel corresponds to the binomial distribution,

$$P_N^{(p)} = \binom{N_{\text{at}}}{N} T_p^N (1 - T_p)^{N_{\text{at}} - N}. \tag{3.9}$$

This is just the probability that out of N_{at} electrons arriving at the barrier N pass through, and others, $N_{\text{at}} - N$, are reflected back. For more than one channel, the binomial distribution does not hold any more: One obtains a convolution of binomial distributions corresponding to each channel. If all transmission eigenvalues are small, (3.9) yields the Poisson distribution, corresponding to the notion of independent electron transfers.

One can now average (3.8) over various distributions of transmission eigenvalues to produce the full counting statistics. In this way, FCS for a double barrier [32], diffusive wires [30, 33], and chaotic cavities [34] were produced.

The concept of FCS can be generalized to different situations, where (3.7) does not apply any more. By now, full counting statistics is a field by itself. Various methods have been developed to calculate higher cumulants, which include the semi-classical approach based on the cascade of Langevin equations, both for diffusive wires [35] and chaotic cavities [36], semi-classical circuit theory based on Keldysh Green's function technique [37–39], semi-classical stochastic path integral method [40], Keldysh sigma-model [41, 42], direct numerical simulation [43], and analytical [44] and numerical [45] treatment of exclusion models. Results on FCS for normal–superconductor interfaces in various situations [37, 46, 47], for charge transfer between superconductors, both for applied constant phase [48] or applied constant voltage [49–52], for quantum to classical crossover in chaotic cavities [53, 54], for FCS in double quantum dots [55], in multi-terminal circuits, including superconducting elements [38, 56], quantum dots in the Coulomb blockade regime [39, 57], interacting diffusive conductors [41, 42, 58, 59], for frequency dependence of the higher cumulants [60–62], and for the time-dependent current [63, 64] are available. There is one issue we would like to mention here. For FCS in the charge transfer between two superconductors with fixed phase difference, the probabilities P_N can sometimes assume negative values [48]. The reason is that phase and charge are canonically conjugated variables and cannot be measured simultaneously. No such problem exists for voltage-biased junctions. Thus, in this case the quantities P_N cannot be interpreted as probabilities. However, Ref. [48] suggested a scheme that makes a measurement of P_N possible, even if they are negative. The set of quantities P_N contains full information about the charge transfer in the system even in this situation.

Let us now specifically consider the third cumulant of the transmitted charge. It has a very important property: At equilibrium ($V = 0$) the characteristic function of (3.7) becomes even in χ, and therefore all the odd cumulants disappear. There is no "Nyquist third cumulant". For this reason, one does not have to measure at very low temperatures to extract the information of transmission eigenvalues. On the other hand, noise measurement is already more complicated than the average current measurement since it requires to collect more measurement results to achieve decent accuracy. The direct measurement of the third cumulant is even more challenging.

From (3.8) we get the ∂ third cumulant in the "shot noise" regime $eV \gg k_B T$,

$$\langle\langle Q^3 \rangle\rangle = e^3 \frac{\partial^3}{\partial (i\chi)^3} \ln \Lambda(\chi) = e^2 V G_Q \Delta t \sum_p T_p(1 - T_p)(1 - 2T_p). \tag{3.10}$$

For a tunnel barrier $T_p \ll 1$ we get $\langle\langle Q^3 \rangle\rangle = e^2 \langle I \rangle \Delta t$, which can be derived directly from the Poisson distribution. In a diffusive wire, the averaging over the distribution function of transmission eigenvalues yields $\langle\langle Q^3 \rangle\rangle = e^2 \langle I \rangle \Delta t / 15$. The third cumulant can be either positive or negative, the open channels with $T_p > 1/2$ favour negative sign. One can also derive the full voltage dependence, including the regime $eV \ll k_B T$, from (3.7).

The first measurement of a higher cumulant of current noise was performed by Reulet, Senzier, and Prober [65], who studied the third cumulant of the voltage drop S_V^3 across a tunnel junction biased by a constant current. The result is plotted

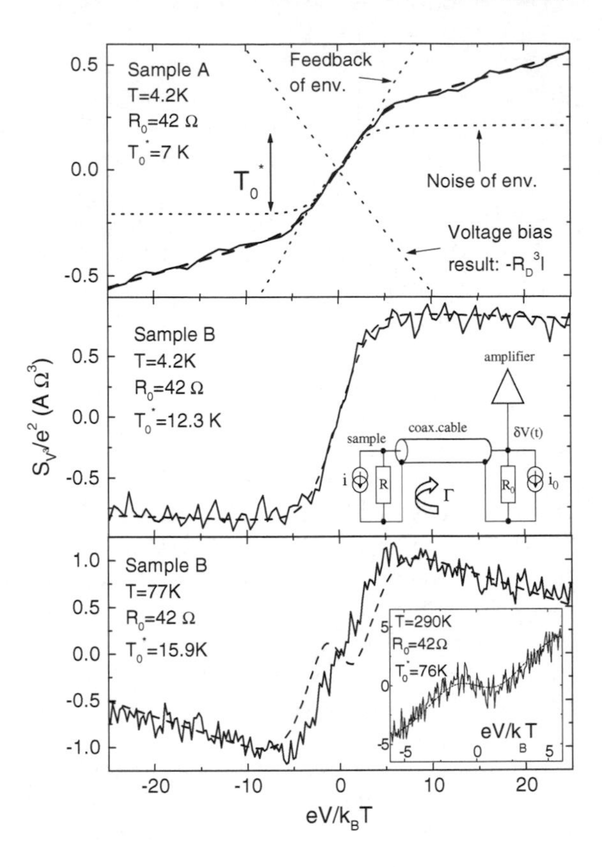

Fig. 3.1 Measurements of the third cumulant [65]. The *top panel* represents comparison of the results with the theory. The *dashed lines* show various theoretical predictions: "naive" multiplication with R_D^3 ("Voltage bias result") and effects of the environment. The agreement is only achieved if the environmental fluctuations are taken into account, see (3.11). The *middle* and the *bottom* panel present measurements on a different sample. The *dashed lines* show the best fits to theoretical predictions. Copyright (2003) by American Physical Society

in Fig. 3.1 against the average voltage drop over the junction (the resistance of the circuit R_D is found independently from the noise measurements). One now has to compare experimental measurements with the theoretical prediction (3.10). Naively, the third cumulant of the voltage is obtained from the third cumulant of the current S_I^3 by multiplication with R_D^3. This operation produces a line shown in the top panel of Fig. 3.1 as "voltage bias result"—in strong contradiction with experimental data. The reason for this disagreement was discovered in Ref. [66]. It is known that, similarly to the average voltage $V = R_D I$, voltage fluctuations can be obtained from the current fluctuations by multiplication with R_D^2, $S_V = R_D^2 S_I$. However, for third and higher cumulants the situation is more complicated, since voltage fluctuations generated at the sample can perturb other parts of the electric circuit and generate there current fluctuations, which affect current fluctuations at the sample. This strongly modifies the relation between current and voltage third cumulants. In the simplest situation, when the environment itself is noiseless ("non-invasive measurement") one has

$$S_{V^3} = -R_D^3 S_{I^3} + 3R_D^4 S_I \frac{dS_I}{dV}. \qquad (3.11)$$

The difference between higher order cumulants is even more dramatic. To characterize voltage fluctuations, one defines the phase $\phi = \int_0^{\Delta t} \delta V(t)dt$ and studies full counting statistics for the phase. It turns out that the phase has Pascal distribution rather than binomial distribution (3.9) which one finds for the transmitted charge; in particular, for the tunnel junction, the distribution is chi square rather than a Poisson one [67, 68].

The problem with the experiment [65] is that the measured third cumulant was dominated by the shot noise contribution (second term in (3.11)), and the contribution S_{I^3} only became important at high temperatures, when the voltage dependence of the noise is weak. To avoid this, several groups performed measurements of the transmitted charge in real time. Following an earlier theoretical suggestion of Ref. [69], Bomze et al. [70] measured the third current cumulant of a tunnel junction by amplifying and analyzing in real time voltage fluctuations on a detector—a resistor with the conductance much higher than that of the sample. The measurements were performed at 4 K and demonstrated the dependence $S_{I^3} = e^2\langle I \rangle$, as expected from the Poisson distribution.

Real-time measurements are easier in the Coulomb blockade regime in quantum dots, since the electron spends a relatively long time in the system, electrons enter the dot one by one, and individual tunnel processes are easier to resolve. The earlier measurements [71] used a single-electron transistor electrostatically coupled to the quantum dot as a detector, and observed real-time detection of single electron tunneling; however, the measurement precision was not enough to extract full counting statistics. Subsequently, more precise real-time detector measurements were performed in quantum dots [72] and in superconducting junction arrays [73], still without extracting the full counting statistics. Then Gustavsson et al. used charge detection with a quantum point contact electrostatically coupled to the quantum dot: When an electron enters the dot, it increases the height of the potential barrier in the quantum point contact. If the detector is tuned close to the step between the plateaus, this increase would block electron passage through the junction. Thus, measurement of the current through the detector gives the real-time information on the occupation of the quantum dot. They plotted the histogram of the transmitted charge during the measurement time Δt (half a second in their experiment) and analyzed the FCS. Very recently Fujisawa et al. [75] reported measurements of FCS in a double quantum dot, also with a quantum point contact as a detector. They collected enough statistics to restore the rates for all possible tunnel processes, checked the detailed balance relation between the rates, produced the occupation probabilities by solving the master equations, and compared the results with the observed FCS.

All these recent experimental advances concentrate on the situations where the FCS from the point of view of the theory is trivial—Poisson distribution in tunnel junctions, and only two possible charge states in quantum dots—and so far serve rather to demonstrate that FCS can be measured. Measurements of less trivial effects, for instance, of tails of the distribution function of the transmitted charge, are still to come.

3.4 Super-Poissonian Noise

It follows from (3.2) that shot noise in the system of non-interacting electrons is always sub-Poissonian: the Fano factor F is less than one. This means that every time super-Poissonian noise is measured the reason must be looked for in interactions (typically electron–electron interactions). However, this statement is too general, and, as many too general statements, useless. Let us look in more detail at different situations which can produce super-Poissonian noise.

As we mentioned, shot noise measures the charge quantum transferred across the nano-structure. An example we already mentioned is an interface between a normal metal and a superconductor. For voltages below the superconducting gap, transport is only possible by Andreev reflection, and the corresponding charge quantum is $2e$. The Fano factor for such a system can be up to $F = 2$. Another example is transport in SNS systems, which proceeds via multiple Andreev reflections. Such process is associated with transfer of Δ/eV charge quanta, which provides super-Poissonian Fano factors.

Let us give another example illustrating the same mechanism. Safonov et al. [76] studied noise in tunneling via localized states. They discovered that the Fano factor strongly depends on the gate voltage, sometimes achieving values above one (Fig. 3.2). To explain these results, Ref. [76] suggested the following model. Imagine the transport occurs through two localized states ("impurities") in parallel: R and M. M is coupled to the leads much more weakly than R, so that its contribution to the current and noise is negligible. If R and M were independent, the Fano factor would vary between 1/2 and 1, depending on the asymmetry of the two barriers separating the impurity from the reservoirs. However, things change if

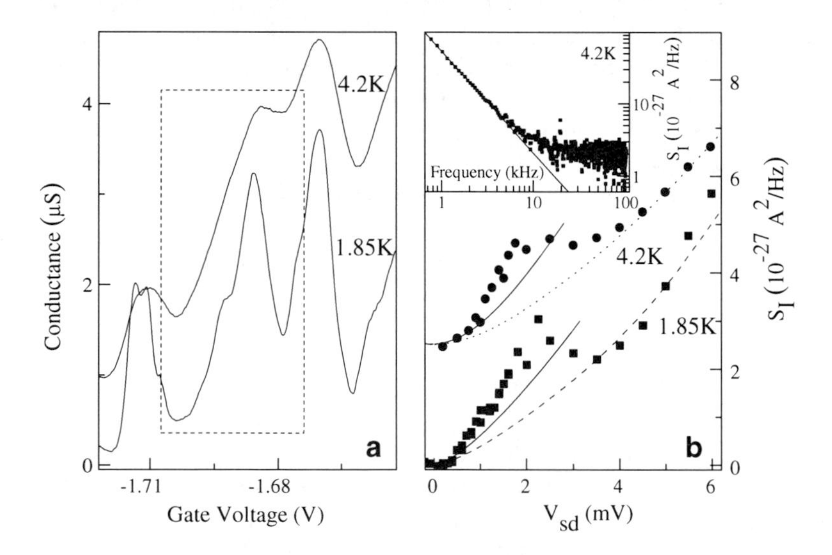

Fig. 3.2 Tunneling via localized states [76]. Noise is shown in the *right panel*; the *solid line* represents the Poisson value. Copyright (2003) by American Physical Society

the two impurities are coupled electrostatically. Due to this coupling, the occupation of M affects the position of electron levels in R and can, for instance, shift a level off the resonance, blocking the transport through R. In this case, if M is occupied, the current through R is blocked, and if M is empty, the current through M proceeds in an ordinary way. Thus, if tunnel rates for M and R are of the order of Γ_R and Γ_M, respectively, transport through the system proceeds in bunches of $\Gamma_R/\Gamma_M \gg 1$ electrons. Thus, the Fano factor can achieve large super-Poissonian values. Since such a two-impurity configuration is not expected to be typical, after impurity averaging, shot noise is considerably reduced, which explains Fano factors slightly above one observed in the experiment.

This example illustrates a general mechanism of super-Poissonian noise. If transport proceeds via two or more electrostatically coupled states, so that occupation of one of the states (M) may block the transport through the other one(s), R, charge is transported in quanta of $e\Gamma_R/\Gamma_M$. Super-Poissonian noise appears provided this number is greater than one, i.e., the states are coupled differently to the leads. This situation may occur in quantum dots in various regimes under the Coulomb blockade condition: One needs that the charging energy is greater than the separation between the energy levels relevant for transport (*dynamical channel blockade*). Theoretical predictions of super-Poissonian noise exist for sequential tunneling regimes in quantum dots with ferromagnetic leads [77–79] (if both leads are partially polarized say spin-up, then spin-up electrons tend to tunnel in bunches, and spin-down electrons block the current for a long time), in a magnetic field [80], general dynamical channel blockade for sequential tunneling [81–84], in double quantum dots [85], and in quantum dots where the level coupling is mediated by non-equilibrium plasmons in the leads [86]. Ref. [87] predicted super-Poissonian noise in the inelastic cotunneling regime.

Recently super-Poissonian noise was experimentally observed by Onac et al. [22] in a carbon nanotube quantum dot (carbon nanotube crossed by two barriers). They measured noise across the Coulomb diamond and found Fano factors up to $F = 3$. Super-Poissonian noise was observed inside the diamonds and is therefore associated with inelastic cotunneling.

Another source of super-Poissonian noise is bistability. An example was provided by Refs. [88, 89], that studied current through quantum wells in the resonant tunneling regime. A similar behavior was observed in tunneling through a zero-dimensional state [90]. Due to interactions, in a certain interval of voltages these wells become bistable: One state with zero current and one state with finite current. For lower voltages, the zero-current branch becomes unstable and ceases to exist; for higher voltages, the finite-current branch does not exist. Noise in such system comes from two sources: (i) “shot” noise—small current fluctuations around each of the branches; (ii) random jumps—random telegraph noise—between the branches. Close to the instability point, when the finite-current branch disappears, current fluctuations around this branch diverge and exceed the Poisson value [2, 91]. Full analysis of the noise also includes large fluctuations, resulting in jumps between the branches [92, 93]. Ref. [94] treats current statistics in a generic bistable system. Experimentally, a link between super-Poissonian noise and bistability is not established convincingly: For instance, a superlattice tunnel diode is a bistable system, but experiments did not discover any super-Poissonian noise enhancement [95].

Finally, we discuss yet another situation, when there are other degrees of freedom in the system which affect transport properties, in particular, noise. Actually, this situation is rather common: In any mesoscopic and macroscopic system, electrons interact with phonons. If this interaction is effective enough—the characteristic length of electron–phonon relaxation L_{ph} is shorter than the size of the system—electrons relax to the equilibrium distribution, and the noise they produce is Nyquist noise—shot noise disappears. If we want to have anything non-trivial, we need to consider non-equilibrium phonons. Such opportunity was recently provided by a new class of devices—nanoelectromechanical systems (NEMS), which couple electron motion to mechanical degrees of freedom. Currently, many species of NEMS were made and investigated, including shuttles—single-electron tunneling devices with a movable central island, double-clamped suspended beams in the Coulomb blockade regime, or single-clamped suspended cantilevers. First, in NEMS one can discuss not only charge noise, but also momentum noise—random fluctuations of momentum transferred from electrons to the crystalline lattice [96–98]. Second, ordinary current noise is strongly modified by mechanical degrees of freedom, both in the shuttling regime [99–102] and for single-electron tunneling [103–108]; in particular, in both situations noise can achieve super-Poisson values. It is not our intention to give here a comprehensive review of noise in NEMS, and we restrict ourself to just one particular situation: a single electron tunneling device weakly coupled to a single-mode underdamped harmonic mechanical oscillator [109, 110].

Qualitatively, the situation is as follows. Imagine that bias and gate voltages are tuned just outside a Coulomb blockade diamond, so that only two charge states are important for electrons: say $n = 0$ and $n = 1$. Non-zero average current means that the number of electrons in the single-electron tunneling device fluctuates randomly between zero and one. The coupling between electrons and vibrations of the oscillator is provided by a force F_n which depends on the charge state and acts on the oscillator. In the regime we discuss this force is a random function that can assume two values, F_0 and F_1. The force generates mechanical oscillations, which in the underdamped case have a large amplitude and a frequency close to the eigenmode ω_0 of the oscillator. The vibration produces feedback on the current since the tunnel rates depend on the position of the oscillator, due to the position dependence of the energy differences available for tunneling. It turns out that the quality factor of the oscillator Q is renormalized due to electron tunneling, but even after the renormalization one still is in the underdamped regime.

Depending on the voltages, one can identify four types for the behavior of the system, which are best described in terms of the probability $P(A)$ to have certain amplitude A of the mechanical oscillator. First, $P(A)$ can be sharply peaked around zero (meaning only very small amplitudes have significant probability) or around a certain finite value. In both cases, noise can be estimated as follows. From dimensional analysis, one obtains $S(0) \sim I^2\tau$, where τ has the dimension of time. In the ordinary situation, τ is of order of the inverse tunnel rate Γ^{-1} (the only time scale in the problem), and one restores the Poisson value of the shot noise. In our case, there is a longer time scale—the decay time $Q/\omega_0 \gg \Gamma^{-1}$; thus, we have $S(0) \sim eI\Gamma Q/\omega_0$, and noise considerably exceeds the Poisson value. In two further cases, the distribution function $P(A)$ has two peaks: either one at $A = 0$ and another one at a finite value of A, or both peaks at finite values of A. This means that only two values of the amplitude are possible. In both cases, on top of the super-Poissonian noise for each peak, we have additional enhancement of noise due to random jumps between the states with different values of the amplitude.

3.5 Interference Effects

Interference effects are at the core of quantum mechanics: They provide information on the phase of the wave function probing its wave nature. These experiments are difficult to perform with electrons in solids. However, we witness steady progress over the last decade, with the number of proposals and successful realisations constantly increasing. We will only discuss here the types of interferometers for which studies of noise are available.

Qualitatively the simplest species is an Aharonov–Bohm (AB) ring—a two-terminal structure where the transmission probability is a periodic function of the magnetic flux penetrating the ring, with the period of the flux quantum. Conductance and shot noise retain this periodic dependence. However, this system has just too few handles, the amplitude of AB oscillations depends essentially on the dynamical phases acquired by electrons moving along the arms of the ring, and the AB vanishes in a ring already with several transport channels.

Recently, Ji et al. [111] performed an experiment with the electronic Mach–Zehnder interferometer (MZI), an analog of the corresponding optical device. The electronic MZI with edge states in the integer quantum Hall regime is shown in Fig. 3.3. In the simplest version, it has one source and two detectors (the voltage V is applied to the source relative to both detectors), the beam splitter A and the beam splitter B, both realized as quantum point contacts. The transport proceeds via the edge states; we assume that in the contact A there is no reflection, so that the edge state proceeds either to the upper or to the lower arm. In the beam splitter B, there is also no reflection, and an electron from either arm can proceed to one of the detectors, 1 or 2. An Aharonov–Bohm flux penetrates the ring. Both conductance between the source and any of the detectors, and current correlations between any reservoirs are flux-dependent. In the experiment, transmission through one of the point contacts could be changed, which creates an additional handle.

Surprisingly, the visibility—the ratio of the phase-dependent and phase-independent parts of the conductance—observed in the experiment, was lower than expected. This suppression of the visibility can be attributed to the loss of phase

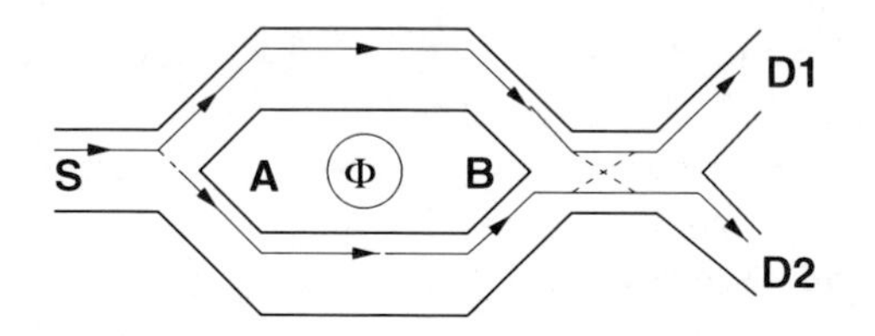

Fig. 3.3 Mach–Zehnder interferometer with edge states. Edge states are shown by *solid lines with arrows*; additional tunnel directions in the beam splitters are indicated by *dashed lines*

coherence, which partially destroys the interference. There could be two reasons for this loss of coherence: phase averaging (for instance, due to the energy dependence of the phase) and dephasing by environment, in particular, by electron-electron interactions. Conductance is affected by both mechanisms in the same way: It is proportional to the transmission probability from the source to the corresponding drain, averaged over the phase. The details depend on the type of the dephasing [112, 113]. Thus, conductance generally cannot distinguish between these two reasons of phase coherence loss, if they have the same dependance on the dephasing time (usually it is proportional to the temperature). It turns out that the situation is different for noise [114, 115], as well as for higher cumulants of the transmitted charge [116]. One has to compare the applied voltage eV with the inverse characteristic lifetime for the phase correlations induced by the environment $\hbar/\tau_c$. In this way, one identifies "fast" $eV\tau_c \ll \hbar$ and "slow" $eV\tau_c \gg \hbar$ environments. For a slow environment, shot noise is merely obtained by averaging of the usual expression, $T(1 - T)$, over the phase. In this case, it does not provide any new information about the environment as compared to the conductance. For fast environments, the behavior of noise is generally different, and thus may provide information about the source of loss of the phase coherence.

In Refs. [117, 118], a more complicated two-particle interferometer was proposed. It consists of four sources and four detectors, separated by four quantum point contacts. Transport is again only possible via the edge states in the integer quantum Hall regime. The setup is designed in such a way that an electron can be transmitted from any source to any detector via *only one* trajectory. The average current is not sensitive to interference, does not depend on the Aharonov–Bohm phase, and is determined by transmission probabilities of the contacts. In contrast, if we consider current cross-correlations at different detectors, the interference contribution originates from the interference of *different* trajectories. In this geometry, these pairs of trajectories are taken in such a way that *together* they enclose a loop and are thus sensitive to the Aharonov–Bohm flux. This is probably up to now the nicest illustration of the two-particle nature of current noise in solid state systems.

Finally, we mention an Andreev interferometer—a normal metal connected by two arms with the same superconducting reservoir. Transport properties of this system are sensitive to magnetic flux enclosed by the arms. Reulet et al. [119]

investigated current noise in this system experimentally and theoretically, and found a periodic dependence on the applied flux. Full counting statistics in an Andreev interferometer was studied in Ref. [120].

Acknowledgments The author very much appreciates collaboration with his friends and colleagues on various issues related to current noise, in chronological order: Stijn van Langen, Eugene Sukhorukov, Henning Schomerus, Carlo Beenakker, Gabriele Campagnano, Oleg Jouravlev, Yuli Nazarov, Omar Usmani, Thomas Ludwig, Alexander Mirlin, and Yuval Gefen. He is especially grateful to Markus Büttiker, who introduced him to the field of shot noise.

References

1. Kogan, Sh.: Electronic Noise and Fluctuations in Solids. Cambridge University Press, Cambridge (1996)
2. Blanter, Ya.M., Büttiker, M.: Phys. Rep. **336**, 1 (2000)
3. Quantum Noise in Mesoscopic Physics: In: Nazarov, Yu.V. (ed.) NATO Science Series II, vol. 97. Dordrecht Boston London, Kluwer (2003)
4. Khlus, V.A.: Zh. Éksp. Teor. Fiz. **93**, 2179 (1987) [Sov. Phys. JETP **66**, 1243 (1987)]
5. Lesovik, G.B.: Pis'ma Zh. Éksp. Teor. Fiz. **49**, 513 (1989) [Sov. Phys. JETP **49**, 592 (1989)]
6. Büttiker, M.: Phys. Rev. Lett. **65**, 2901 (1990)
7. Landauer, R., Martin, T.: Physica B **167**, 175 (1991) [ibid 182, 288(E) (1992)]
8. Oberholzer, S., Sukhorukov, E.V., Strunk, C., Schönenberger, C., Heinzel, T., Holland, M.: Phys. Rev. Lett. **86**, 2114 (2001)
9. Oberholzer, S., Sukhorukov, E.V., Schönenberger, C.: Nature **415**, 765 (2002)
10. Jehl, X., Sanquer, M., Calemczuk, R., Mailly, D.: Nature **405**, 50 (2000)
11. Kozhevnikov, A.A., Schoelkopf, R.J., Prober, D.E.: Phys. Rev. Lett. **84**, 3398 (2000)
12. Cron, R., Goffman, M.F., Esteve, D., Urbina, C.: Phys. Rev. Lett. **86**, 4104 (2001)
13. Lefloch, F., Hoffmann, C., Sanquer, M., Quirion, D.: Phys. Rev. Lett. **90**, 067002 (2003)
14. Hoffmann, C., Lefloch, F., Sanquer, M., Pannetier, B.: Phys. Rev. B **70**, 180503 (2004)
15. Camino, F.E., Kuznetsov, V.V., Mendez, E.E., Schäpers, Th., Guzenko, V.A., Hardtdegen, H.: Phys. Rev. B **71**, 020506 (2005)
16. Nauen, A., Hapke-Wurst, I., Hohls, F., Zeitler, U., Haug, R.J., Pierz, K.: Phys. Rev. B **66**, 161303 (2002)
17. Oberholzer, S., Sukhorukov, E.V., Strunk, C., Schönenberger, C.: Phys. Rev. B **66**, 233304 (2002)
18. Kuznetsov, V.V., Mendez, E.E., Zuo, X., Snider, G.L., Croke, E.T.: Phys. Rev. Lett. **85**, 397 (2000)
19. Camino, F.E., Kuznetsov, V.V., Mendez, E.E., Gershenson, M.E., Reuter, D., Schafmeister, P., Wieck, A.D.: Phys. Rev. B **68**, 073313 (2003)
20. Reydellet, L.-H., Roche, P., Glattli, D.C., Etienne, B., Jin, Y.: Phys. Rev. Lett. **90**, 176803 (2003)
21. Nauen, A., Hohls, F., Maire, N., Pierz, K., Haug, R.J.: Phys. Rev. B **70**, 033305 (2004)
22. Onac, E., Balestro, F., Trauzettel, B., Lodewijk, C.F.J., Kouwenhoven, L.P.: Phys. Rev. Lett. **96**, 026803 (2006)
23. Aguado, R., Kouwenhoven, L.P.: Phys. Rev. Lett. **84**, 1986 (2000)
24. Schoelkopf, R.J., Clerk, A.A., Girvin, S.M., Lehnert, K.W., Devoret, M.H.: In Quantum Noise in Mesoscopic Physics: In: Nazarov, Yu.V. (ed.) NATO Science Series II, vol. 97. Dordrecht Boston London, Kluwer (2003), available also as cond-mat/0210247
25. Deblock, R., Onac, E., Gurevich, L., Kouwenhoven, L.P.: Science **301**, 203 (2003)

26. Onac, E., Deblock, R., Kouwenhoven, L.P.: unpublished; also, Onac, E.: High-frequency noise detection in mesoscopic devices. PhD Thesis, Delft University of Technology, Delft (2005)
27. Choi, M.-S., Plastina, F., Fazio, R.: Phys. Rev. B **67**, 045105 (2003)
28. Onac, E., Balestro, F., Hartmann, U., Nazarov, Yu.V., Kouwenhoven, L.P.: unpublished; also, Onac, E.: High-frequency noise detection in mesoscopic devices. PhD Thesis, Delft University of Technology, Delft (2005)
29. Levitov, L.S., Lesovik, G.B.: Pis'ma Zh. Éksp. Teor. Fiz. **58**, 225 (1993) [JETP Lett. **58**, 230 (1993)]
30. Lee, H., Levitov, L.S., Yakovets, A.Yu.: Phys. Rev. B **51**, 4079 (1995)
31. Levitov, L.S., Lee, H., Lesovik, G.B.: J. Math. Phys. **37**, 4845 (1996)
32. de Jong, M.J.M.: Phys. Rev. B **54**, 8144 (1996)
33. Nazarov, Yu.V.: Ann. Phys. (Leipzig) **8**, 507 (1999)
34. Blanter, Ya.M., Schomerus, H., Beenakker, C.W.J.: Physica E **11**, 1 (2001)
35. Nagaev, K.E.: Phys. Rev. B **66**, 075334 (2002)
36. Nagaev, K.E., Samuelsson, P., Pilgram, S.: Phys. Rev. B **66**, 195318 (2002)
37. Belzig, W., Nazarov, Yu.V.: Phys. Rev. Lett. **87**, 067006 (2001)
38. Nazarov, Yu.V., Bagrets, D.A.: Phys. Rev. Lett. **88**, 196801 (2002)
39. Bagrets, D.A., Nazarov, Yu.V.: Phys. Rev. Lett. **67**, 085316 (2003)
40. Pilgram, S., Jordan, A.N., Sukhorukov, E.V., Büttiker, M.: Phys. Rev. Lett. **90**, 206801 (2003)
41. Gutman, D.B., Gefen, Y.: Phys. Rev. B **68**, 035302 (2003)
42. Gutman, D.B., Mirlin, A.D., Gefen, Y.: Phys. Rev. B **71**, 085118 (2005)
43. Pala, M.G., Iannaccone, G.: Phys. Rev. Lett. **93**, 256803 (2004)
44. Roche, P.-E., Douçot, B.: Eur. Phys. J. B **27**, 393 (2002)
45. Roche, P.-E., Derrida, B., Douçot, B.: Eur. Phys. J. B **43**, 529 (2005)
46. Muzykantskii, B.A., Khmelnitskii, D.E.: Phys. Rev. B **50**, 3982 (1994)
47. Samuelsson, P.: Phys. Rev. B **67**, 054508 (2003)
48. Belzig, W., Nazarov, Yu.V.: Phys. Rev. Lett. **87**, 197006 (2001)
49. Belzig, W., Samuelsson, P.: Europhys. Lett. **64**, 253 (2003)
50. Cuevas, J.C., Belzig, W.: Phys. Rev. Lett. **91**, 187001 (2003)
51. Johansson, G., Samuelsson, P., Ingerman, Å.: Phys. Rev. Lett. **91**, 187002 (2003)
52. Cuevas, J.C., Belzig, W.: Phys. Rev. B **70**, 214512 (2004)
53. Whitney, R.S., Jacquod, Ph.: Phys. Rev. Lett. **94**, 116801 (2005)
54. Sukhorukov, E.V., Bulashenko, O.M.: Phys. Rev. Lett. **94**, 116803 (2005)
55. Kießlich, G., Samuelsson, P., Wacker, A., Schöll, E.: Phy. Rev. B **73**, 033312 (2006)
56. Börlin, J., Belzig, W., Bruder, C.: Phys. Rev. Lett. **88**, 197001 (2002)
57. Bagrets, D.A., Nazarov, V.Yu.: Phys. Rev. Lett. **94**, 056801 (2005)
58. Pilgram, S.: Phys. Rev. B **69**, 115315 (2004)
59. Bagrets, D.A.: Phys. Rev. Lett. **93**, 236803 (2004)
60. Galaktionov, A.V., Golubev, D.S., Zaikin, A.D.: Phys. Rev. Lett. **68**, 235333 (2003)
61. Nagaev, K.E., Pilgram, S., Büttiker, M.: Phys. Rev. Lett. **92**, 176804 (2004)
62. Pilgram, S., Nagaev, K.E., Büttiker, M.: Phys. Rev. B **70**, 045304 (2004)
63. Ivanov, D.A., Levitov, L.S.: Pis'ma Zh. Éksp. Teor. Fiz. **58**, 450 (1993) [JETP Lett. **58**, 461 (1993)]
64. Ivanov, D.A., Lee, H.W., Levitov, L.S.: Phys. Rev. B **56**, 6839 (1997)
65. Reulet, B., Senzier, J., Prober, D.E.: Phys. Rev. Lett. **91**, 196601 (2003)
66. Beenakker, C.W.J., Kindermann, M., Nazarov, Yu.V.: Phys. Rev. Lett. **90**, 176802 (2003)
67. Kindermann, M., Nazarov, Yu.V., Beenakker, C.W.J.: Phys. Rev. Lett. **90**, 246805 (2003)
68. Kindermann, M., Nazarov, Yu.V., Beenakker, C.W.J.: Phys. Rev. B **69**, 035336 (2004)
69. Levitov, L.S., Reznikov, M.: Phys. Rev. B **70**, 115305 (2004)
70. Bomze, Yu., Gershon, G., Shovkun, D., Levitov, L.S., Reznikov, M.: Phys. Rev. Lett. **95**, 176601 (2005)

71. Lu, W., Ji, Z., Pfeiffer, L., West, K.W., Rimberg, A.J.: Nature **423**, 422 (2003)
72. Fujisawa, T., Hayashi, T., Hirayama, Y., Cheong, H.D., Jeong, Y.H.: Appl. Phys. Lett. **84**, 2343 (2004)
73. Bylander, J., Duty, T., Delsing, P.: Nature **434**, 361 (2005)
74. Gustavsson, S., Leturcq, R., Simovič, B., Schleser, R., Ihn, T., Studerus, P., Ensslin, K., Driscoll, D.C., Gossard, A.C.: Phys. Rev. Lett. **96**, 076605 (2006)
75. Fujisawa, T., Hayashi, T., Tomita, R., Hirayama, Y.: Science **312**, 1634 (2006)
76. Safonov, S.S., Savchenko, A.K., Bagrets, D.A., Jouravlev, O.N., Nazarov, Yu.V., Linfield, E.H., Ritchie, D.A.: Phys. Rev. Lett. **91**, 136801 (2003)
77. Bułka, B.R.: Phys. Rev. B **62**, 1186 (2000)
78. Cottet, A., Belzig, W., Bruder, C.: Phys. Rev. Lett. **92**, 206801 (2004)
79. Braun, M., König, J., Martinek, J.: unpublished
80. Cottet, A., Belzig, W.: Europhys. Lett. **66**, 405 (2004)
81. Kießlich, G., Wacker, A., Schöll, E.: Phys. Rev. B **68**, 125320 (2003)
82. Cottet, A., Belzig, W., Bruder, C.: Phys. Rev. B **70**, 115315 (2004)
83. Thielmann, A., Hettler, M.H., König, J., Schön, G.: Phys. Rev. B **71**, 045341 (2005)
84. Belzig. W.: Phys. Rev. B **71**, 161301 (2005)
85. Djuric, I., Dong, B., Cui, H.L.: Appl. Phys. Lett. **87**, 032105 (2005)
86. Kim, J.U., Kinaret, J.M., Choi, M.S.: J. Phys. Condensed Matter **17**, 3815 (2005)
87. Sukhorukov, E.V., Burkard, G., Loss, D.: Phys. Rev. B **63**, 125315 (2001)
88. Iannaccone, G., Lombardi, G., Macucci, M., Pellegrini, B.: Phys. Rev. Lett. **80**, 1054 (1998)
89. Kuznetsov, V.V., Mendez, E.E., Bruno, J.D., Pham, J.T.: Phys. Rev. B **58**, 10159 (1998)
90. Nauen, A., Hohls, F., Konemann, J., Haug, R.J.: Phys. Rev. B **69**, 113316 (2004)
91. Blanter, Ya.M., Büttiker, M.: Phys. Rev. B **59**, 10217 (1999)
92. Tretiakov, O.A., Gramespacher, T., Matveev, K.A.: Phys. Rev. B **67**, 073303 (2003)
93. Tretiakov, O.A., Matveev, K.A.: Phys. Rev. B **71**, 165326 (2005)
94. Jordan, A.N., Sukhorukov, E.V.: Phys. Rev. Lett. **93**, 260604 (2004); ibid 94:059901(E) (2005)
95. Song, W., Mendez, E.E., Kuznetsov, V., Nielsen, B.: Appl. Phys. Lett. **82**, 1568 (2003)
96. Shytov, A.V., Levitov, L.S., Beenakker, C.W.J.: Phys. Rev. Lett. **88**, 228303 (2002)
97. Kindermann, M., Beenakker, C.W.J.: Phys. Rev. B **66**, 224106 (2002)
98. Tajic, A., Kindermann, M., Beenakker, C.W.J.: Phys. Rev. B **66**, 241301 (2002)
99. Novotný, T., Donarini, A., Flindt, C., Jauho, A.-P.: Phys. Rev. Lett. **92**, 248302 (2004)
100. Isacsson, A., Nord, T.: Europhys. Lett. **66**, 708 (2004)
101. Pistolesi, F.: Phys. Rev. B **69**, 245409 (2004)
102. Flindt, C., Novotný, T., Jauho, A.-P.: Phys. Rev. B **70**, 205334 (2004)
103. Zhu, J.X., Balatsky, A.V.: Phys. Rev. B **67**, 165326 (2003)
104. Armour, A.D.: Phys. Rev. B **70**, 165315 (2004)
105. Chtchelkachtchev, N.M., Belzig, W., Bruder, C.: Phys. Rev. B **70**, 193305 (2004)
106. Koch, J., Raikh, M.E., von Oppen, F.: Phys. Rev. Lett. **95**, 056801 (2005)
107. Yu, H., Liang, J.Q.: Phys. Rev. B **72**, 075371 (2005)
108. Rodriguez, D.A., Armour, A.D.: Phys. Rev. B **72**, 085324 (2005)
109. Blanter, Ya.M., Usmani, O., Nazarov, Yu.V.: Phys. Rev. Lett. **93**, 136802 (2004); ibid 94:049904(E) (2005)
110. Usmani, O., Blanter, Ya.M., Nazarov, Yu.V.: Phys. Rev. B **75**, 195312 (2007)
111. Ji, Y., Chung, Y., Sprinzak, D., Heiblum, M., Mahalu, D., Shtrikman, H.: Nature **422**, 415 (2003)
112. Seelig, G., Büttiker, M.: Phys. Rev. B **64**, 245313 (2001)
113. Seelig, G., Pilgram, S., Jordan, A.N., Büttiker, M.: Phys. Rev. B **68**, 161310 (2003)
114. Marquardt, F., Bruder, C.: Phys. Rev. Lett. **92**, 056805 (2004)
115. Marquardt, F., Bruder, C.: Phys. Rev. B **70**, 125305 (2004)
116. Förster, H., Pilgram, S., Büttiker, M.: Phys. Rev. B **72**, 075301 (2005)

117. Samuelsson, P., Sukhorukov, E.V., Büttiker, M.: Phys. Rev. Lett. **92**, 026805 (2004)
118. Chung, V.S.-W., Samuelsson, P., Büttiker, M.: Phys. Rev. B **72**, 125320 (2005)
119. Reulet, B., Kozhevnikov, A.A., Prober, D.E., Belzig, W., Nazarov, Yu.V.: Phys. Rev. Lett. **90**, 066601 (2003)
120. Bezuglyi, E.V., Bratus', E.N., Shumeiko, V.S., Vinokur, V.: Phys. Rev. B **70**, 064507 (2004)

Chapter 4
Josephson Qubits as Probes of 1/*f* Noise

Alexander Shnirman, Gerd Schön, Ivar Martin and Yuriy Makhlin

4.1 Introduction

Josephson junction based systems are one of the promising candidates for quantum state engineering with solid state systems. In recent years great progress was achieved in this area. After initial breakthroughs of the groups in Saclay and NEC (Tsukuba) in the late 1990s, there are now many experimental groups worldwide working in this area, many of them with considerable previous experience in nano-electronics. By now the full scope of single-qubit (NMR-like) control is possible. One can drive Rabi oscillations, observe Ramsey fringes, apply composite pulses and echo technique [1]. The goal of "single shot" measurements has almost been achieved [2, 3]. There are first reports about 2-bit operations [4]. The decoherence times have reached microseconds, which would allow for hundreds of gates. Finally, a setup equivalent to cavity QED was realized in superconducting circuits [5]. We refer the reader to the recent reviews [6, 7].

Despite the great progress decoherence remains the limiting factor in solid state circuits. Since one wants to manipulate and measure the qubits, some decoherence is unavoidable. There are, however, noise sources which are purely intrinsic, i.e., they are not related to any controlling or measuring circuitry. Eliminating those sources as much as possible is therefore of greatest importance. The main intrinsic

A. Shnirman (✉) and G. Schön
Institut für Theoretische Festkörperphysik, Karlsruhe Institute of Technology (KIT), 76128 Karlsruhe, Germany
e-mail: alexander.shnirman@kit.edu

I. Martin
Theoretical Division, Los Alamos National Laboratory, Los Alamos, 87545 NM, USA

Y. Makhlin
Landau Institute for Theoretical Physics, Kosygin st. 2, 119334 Moscow, Russia

M. Vojta et al. (eds.), *CFN Lectures on Functional Nanostructures – Volume 2*,
Lecture Notes in Physics 820, DOI: 10.1007/978-3-642-14376-2_4,

source of decoherence in most superconducting qubits is $1/f$ noise of either the charge, the flux, or the critical Josephson current.

On the other hand, the full control of 1-qubit circuits opens the possibility to use qubits as efficient noise detectors [8, 9]. The idea is to measure the decoherence times of the qubit while changing its parameters and extract from the data the noise in the qubit's environment. An experiment of this type was performed by Astafiev et al. [10]. Further information about the noise was obtained in recent studies [11, 12]. In this paper we give a short overview of new and improved understanding of the nature of $1/f$ noise.

4.2 Charge Qubit and Charge Noise

To introduce the basic concepts we consider the simplest charge qubit. The system is shown in Fig. 4.1. Its Hamiltonian reads

$$H = \sum_n \left[E_{\text{ch}}(n, V_g)|n\rangle\langle n| + \frac{E_J}{2}|n\rangle\langle n \pm 1| \right], \tag{4.1}$$

where n is the number of Cooper pairs on the island, with the charging energy given by

$$E_{\text{ch}}(n, V_g) = \frac{(2ne - Q_g)^2}{2(C_g + C_J)}, \tag{4.2}$$

and the induced gate charge is $Q_g = C_g V_g$. Near $Q_g = e$ one can consider the two lowest energy charge states. In the spin $- 1/2$ representation one obtains the following Hamiltonian

$$H = -\frac{1}{2}\Delta E_{\text{ch}}(V_g)\,\hat{\sigma}_z - \frac{1}{2}E_J\,\hat{\sigma}_x, \tag{4.3}$$

Introducing an angle $\eta(V_g)$ such that $\tan\eta = E_J/E_{\text{ch}}(V_g)$ we rewrite the Hamiltonian as

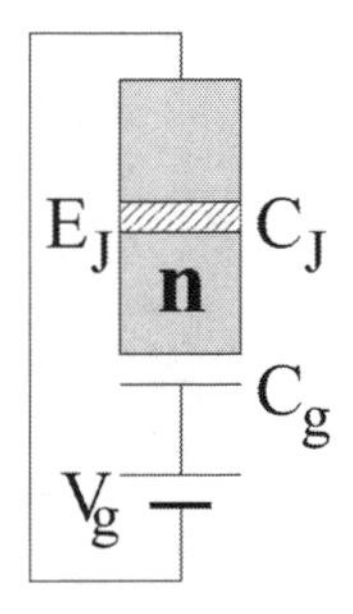

Fig. 4.1 Charge qubit: The two *gray areas* are superconducting islands, with a Josephson coupling E_J. n denotes the number of Cooper pairs on the lower island

$$H = -\frac{1}{2}\Delta E\,(\cos\eta\,\hat{\sigma}_z + \sin\eta\,\hat{\sigma}_x). \tag{4.4}$$

We now assume that the gate charge has a noisy component, i.e., $Q_g = C_g V_g + \delta Q$. Then the charging energy fluctuates and we obtain

$$H = -\frac{1}{2}\Delta E\,(\cos\eta\,\hat{\sigma}_z + \sin\eta\,\hat{\sigma}_x) - \frac{1}{2}X\hat{\sigma}_z, \tag{4.5}$$

where $X = e\delta Q/(C_g + C_J)$. In the eigenbasis of the qubit this gives

$$H = -\frac{1}{2}\Delta E\,\hat{\sigma}_z - \frac{1}{2}X(\cos\eta\,\hat{\sigma}_z - \sin\eta\,\hat{\sigma}_x). \tag{4.6}$$

For sufficiently weak noise with regular spectrum $S_X(\omega)$, the Bloch–Redfield theory [13, 14] gives the dissipative rates. The relaxation (spin flip) rate is given by

$$\Gamma_1 \equiv \frac{1}{T_1} = \frac{1}{2}\sin^2\eta\;S_X(\omega = \Delta E), \tag{4.7}$$

while the dephasing rate

$$\Gamma_2 \equiv \frac{1}{T_2} = \frac{1}{2}\Gamma_1 + \Gamma_\varphi, \tag{4.8}$$

with

$$\Gamma_\varphi = \frac{1}{2}\cos^2\eta S_X(\omega = 0). \tag{4.9}$$

is a combination of spin-flip effects (Γ_1) and of the so called "pure" dephasing, characterized by the rate $\Gamma_\varphi = 1/T_2^*$. The pure dephasing is usually associated with the inhomogeneous level broadening in ensembles of spins, but occurs also for a single spin due to the "longitudinal" (coupling to σ_z) low-frequency noise.

We now consider the situation where the noise X is characterized by the spectral density

$$S_X(\omega) = \frac{\alpha}{|\omega|} \tag{4.10}$$

in the interval of frequencies $\omega_{\mathrm{ir}} < \omega < \omega_{\mathrm{c}}$. In this case (4.9) is clearly inapplicable. Several models of $1/f$ noise and pure dephasing were developed in the literature [15–18]. In all of them the T_2-decay of the coherences (i.e., of the off-diagonal elements of the density matrix) is given by decay law $e^{-\Gamma_1 t/2}f(t)$. The pure decoherence described by the function $f(t)$ depends on the statistics of the noise. For our purposes here a very rough estimate is enough. When deriving the Bloch–Redfield results, e.g., (4.9), one realizes that $S(\omega = 0)$ should actually be understood as the noise power averaged over the frequency band of width $\sim\Gamma_\varphi$ around

$\omega = 0$. We, thus, obtain a time scale of the pure dephasing from the self-consistency condition $\Gamma_\varphi = S_X(\Gamma_\varphi)$. This gives

$$\Gamma_\varphi \approx \sqrt{\alpha} \cos \eta. \tag{4.11}$$

For the cases of "strongly non-Gaussian" statistics [18], α and Γ_φ should be understood as typical rather than ensemble averaged quantities. From the study of many examples we came to the conclusion that the relation (4.11) is universal irrespective of the noise statistics as long as $\Gamma_\varphi > \omega_{\text{ir}}$.

4.3 Analysis of the NEC Experiments

Astafiev et al. [10] measured the T_1 and T_2^* time scales in a charge qubit. As the energy splitting ΔE and the angle η were independently controlled, they could extract the noise power $S(\omega)$ in the GHz range using (4.7). In addition they were able to determine the strength of the $1/f$ noise, α, using (4.11). The results suggested a connection between the strengths of the Ohmic high-frequency noise, responsible for the relaxation of the qubit (T_1-decay), and the low-frequency $1/f$ noise, which dominates the dephasing (T_2-decay). The noise power spectra, extrapolated from the low- and high-frequency sides, turn out to cross at ω of order T. Expressing the high-frequency noise at $\omega > T$ as $S_X(\omega) = a\omega$, they found that the strength of the low-frequency noise scales as $\alpha = aT^2$ (see Fig. 4.2). The T^2 dependence of the low-frequency noise power was observed earlier for the $1/f$noise in Josephson devices [19, 20]. Further evidence for the T^2 behavior was obtained recently [21, 22]. But the fact that the two parts of the spectrum are characterized by the same constant a was surprising.

4.4 Resonances in Phase Qubits

Additional information was obtained from experiments with phase qubits (current biased large area Josephson junction) by Simmonds et al. [23]. These experiments

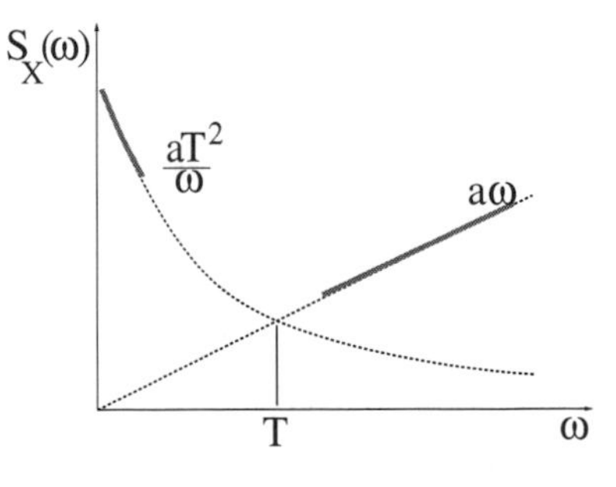

Fig. 4.2 Asymptotic behavior of noise at low and high frequencies

revealed the presence of spurious quantum two-level systems with strong effects on the high-frequency (~ 10 GHz) qubit dynamics. In a phase qubit one controls the energy splitting between the qubit states by changing the bias current. It turned out that at certain values of the bias current the system ceased to behave as a two-level system but showed rather a 4-level dynamics. This phenomenon can be attributed to the existence of a collection of coherent two-level fluctuators (TLF) in the oxide of the tunnel barrier. When the energy splitting of the qubit coincides with that of one of the fluctuators a pair of states $|g_{\text{qubit}}\rangle|e_{\text{fluctuator}}\rangle$ and $|e_{\text{qubit}}\rangle|g_{\text{fluctuator}}\rangle$ are degenerate. This degeneracy is lifted by the qubit–fluctuator interaction, which leads to a gap (avoided crossing) in the spectroscopy of the qubit. There are at least two types of interactions which could be responsible for lifting the degeneracy [24]. One corresponds to a situation in which a two-level fluctuator blocks a conducting channel in one of its states and, thus, influences the Josephson energy of the junction [23]. The other arises due to a dipole moment of the fluctuator interacting with the electric field in the junction. Recent studies [12] point towards the second option. The most surprising was the observation that the two-level fluctuators are more coherent than the qubit. Hence, the decoherence of the 4-level system in a resonant situation is dominated by the decoherence of the qubit.

4.5 High- and Low-Frequency Noise from Coherent TLFs

Motivated by the above mentioned experiments we have pointed out [25] that a set of *coherent* two-level systems may produce both high- and low-frequency noise with strengths that are naturally related. As a model we consider a set of coherent two-level systems described by the Pauli matrices $\sigma_{p,j}$, where $p = x, y, z$, and j labels the particular TLF. We write the Hamiltonian of the set in the basis such that their contributions to the relevant fluctuating quantity, e.g., the gate charge, are $X \equiv \sum_j v_j \sigma_{z,j}$. Then

$$H_{\text{TLS}} = \sum_j \left[-\frac{1}{2}(\varepsilon_j \sigma_{z,j} + \Delta_j \sigma_{x,j}) + H_{\text{diss},j} \right]. \tag{4.12}$$

Here, in the language of tunneling TLSs (TTLS), ε_j are the bias energies and Δ_j the tunnel amplitudes between two states. Each individual TLS, j, is subject to dissipation due to its own bath with Hamiltonian $H_{\text{diss},j}$. We do not specify $H_{\text{diss},j}$, but only assume that it produces the usual relaxation (T_1) and dephasing (T_2) processes. We assume that all the TLSs are under-damped, with $\Gamma_{1,j} \equiv T_{1,j}^{-1} \ll E_j$ and $\Gamma_{2,j} \equiv T_{2,j}^{-1} \ll E_j$. Here $E_j \equiv \sqrt{\epsilon_j^2 + \Delta_j^2}$ is the energy splitting.

Our goal in the following is to investigate the noise properties of the fluctuating field X. For that reason we evaluate the (unsymmetrized) correlator

$$C_X(\omega) \equiv \int dt \Big\{ \langle X(t)X(0)\rangle - \langle X\rangle^2 \Big\} e^{i\omega t}. \tag{4.13}$$

For independent TLSs the noise is a sum of individual contributions, $C_X = \sum_j v_j^2 C_j$, where

$$C_j(\omega) \equiv \int dt \Big\{ \langle \sigma_{z,j}(t)\sigma_{z,j}(0)\rangle - \langle \sigma_{z,j}\rangle^2 \Big\} e^{i\omega t}. \tag{4.14}$$

To obtain C_j we first transform to the eigenbasis of the TLS. This gives

$$H_{\mathrm{TLS}} = \sum_j \Big\{ -\frac{1}{2} E_j \rho_{z,j} + H_{\mathrm{diss},j} \Big\}, \tag{4.15}$$

and

$$X = \sum_j v_j \left(\cos\theta_j\, \rho_{z,j} - \sin\theta_j\, \rho_{x,j}\right), \tag{4.16}$$

where $\tan\theta_j \equiv \Delta_j/\epsilon_j$. Proceeding in the spirit of the Bloch–Redfield theory [13, 14] we readily find

$$\begin{aligned} C_j(\omega) \approx{}& \cos^2\theta_j \left[1 - \langle \rho_{z,j}\rangle^2\right] \frac{2\Gamma_{1,j}}{\Gamma_{1,j}^2 + \omega^2} \\ &+ \sin^2\theta_j \left[\frac{1 + \langle \rho_{z,j}\rangle}{2}\right] \frac{2\Gamma_{2,j}}{\Gamma_{2,j}^2 + (\omega - E_j)^2} \\ &+ \sin^2\theta_j \left[\frac{1 - \langle \rho_{z,j}\rangle}{2}\right] \frac{2\Gamma_{2,j}}{\Gamma_{2,j}^2 + (\omega + E_j)^2}. \end{aligned} \tag{4.17}$$

In thermal equilibrium we have $\langle \rho_{z,j}\rangle = \tanh(E_j/2T)$. The first term, due to the longitudinal part of the coupling, describes random telegraph noise of a thermally excited TLS. We have assumed $\Gamma_{1,j} \ll T$, so that this term is symmetric (classical). The second term is due to the transverse coupling and describes absorption by the TLS, while the third term describes the transitions of the TLS with emission. We observe that TLSs with $E_j \gg T$ contribute to C_X only at the (positive) frequency $\omega = E_j$. Indeed their contribution at $\omega = 0$ is suppressed by the thermal factor $1 - \langle \rho_{z,j}\rangle^2 = 1 - \tanh^2(E_j/2T)$. Also the negative frequency (emission) contribution at $\omega = -E_j$ is suppressed. These high-energy TLSs remain always in their ground state. Only the TLSs with $E_j < T$ are thermally excited, performing real random transitions between their two eigenstates, and contribute at $\omega = \pm E_j$ and at $\omega = 0$. Note that the separation of the terms in (4.17) into low- and high-frequency noise is meaningful only provided the typical width $\Gamma_{1,j}$ of the low-ω

Lorentzians is lower than the high frequencies of interest, which are defined, e.g., by the qubit's level splitting or temperature.

For a dense distribution of the parameters ϵ, Δ, and v we can evaluate the low- and high-frequency noise. For positive high frequencies, $\omega \gg T$, we obtain

$$\begin{aligned} C_X(\omega) &\approx \sum_j v_j^2 \sin^2\theta_j \frac{2\Gamma_{2,j}}{\Gamma_{2,j}^2 + (\omega - E_j)^2} \\ &\approx N \int d\epsilon\, d\Delta\, dv\, P(\epsilon, \Delta, v) v^2 \sin^2\theta \cdot 2\pi\delta(\omega - E), \end{aligned} \tag{4.18}$$

where N is the number of fluctuators, $P(\varepsilon, \Delta, v)$ is the distribution function normalized to 1, $E \equiv \sqrt{\epsilon^2 + \Delta^2}$, and $\tan\theta = \Delta/\epsilon$. Without loss of generality we take $\epsilon \geq 0$ and $\Delta \geq 0$.

At negative high frequencies ($\omega < 0$ and $|\omega| > T$) the correlator $C_X(\omega)$ is exponentially suppressed. On the other hand, the total weight of the low-frequency noise (up to $\omega \approx \Gamma_{1,\max}$, where $\Gamma_{1,\max}$ is the maximum relaxation rate of the TLSs) follows from the first term of (4.17). (Since we have assumed $\Gamma_{1,j} \ll E_j$ we can disregard the contribution of the last two terms of (4.17). Each Lorentzian contributes 1. Thus we obtain

$$\begin{aligned} &\int_{\text{low freq.}} \frac{d\omega}{2\pi} C_X(\omega) \\ &\approx \int_{\text{low freq.}} \frac{d\omega}{2\pi} \sum_j v_j^2 \cos^2\theta_j \left[1 - \langle\rho_{z,j}\rangle^2\right] \frac{2\Gamma_{1,j}}{\Gamma_{1,j}^2 + \omega^2} \\ &\approx N \int d\epsilon\, d\Delta\, dv\, P(\epsilon, \Delta, v)\, v^2 \cos^2\theta \,\frac{1}{\cosh^2 \frac{E}{2T}}. \end{aligned} \tag{4.19}$$

Equations 4.18 and 4.19 provide the general framework for further discussion.

Next we investigate possible distributions for the parameters ϵ, Δ, and v. We consider a log-uniform distribution of tunnel splittings Δ, with density $P_\Delta(\Delta) \propto 1/\Delta$ in a range $[\Delta_{\min}, \Delta_{\max}]$. This distribution is well known to provide for the $1/f$ behavior of the low-frequency noise [26]. It is natural for TTLSs as Δ is an exponential function of, e.g., the tunnel barrier height [27], which is an almost uniformly distributed parameter. The relaxation rates are, then, also distributed log-uniformly, $P_{\Gamma_1}(\Gamma_1) \propto 1/\Gamma_1$, and the sum of many Lorentzians of width Γ_1 centered at $\omega = 0$ adds up to the $1/f$ noise.

The distribution of v is rather arbitrary. We only assume that it is uncorrelated with ε and Δ. Finally we have to specify the distribution of ϵ. First, we assume that the temperature is lower than $\Delta_{\max}$. For the high-frequency part, $T < \omega < \Delta_{\max}$, we find, after taking the integral over Δ in (4.18), that

$$C_X(\omega) \propto \frac{1}{\omega} \int_0^{\omega} P_\varepsilon(\varepsilon) d\varepsilon \,. \tag{4.20}$$

This is consistent with the observed Ohmic behavior $C_X \propto \omega$ only for a linear distribution $P_\varepsilon(\varepsilon) \propto \varepsilon$.

Remarkably, this distribution, $P(\varepsilon, \Delta) \propto \varepsilon/\Delta$, produces at the same time the $T^2 \ln(T/\Delta_{\rm min})$ behavior of the low-frequency weight (4.19), observed in several experiments [19–22]. If the low-frequency noise has a $1/f$ dependence, the two parts of the spectrum would cross around $\omega \sim T$ [10].

In the opposite limit, $T \gg \Delta_{\rm max}$, the high-frequency noise depends on the detailed shape of the cutoff of $P_\Delta(\Delta)$ at $\Delta_{\rm max}$. As an example, for a hard cutoff the Ohmic spectral density implies that $P_\varepsilon \propto \varepsilon^3$, and the low-frequency weight scales with T^4. For a $1/f$ low-frequency behavior, the spectra would cross at $\omega \sim T^2/\Delta_{\rm max} \gg T$, which is not in agreement with the result of Ref. [10].

A remark is in order concerning the crossing at $\omega \approx T$ discussed above. It is not guaranteed that the spectrum has a $1/f$ dependence up to $\omega \sim T$. Rather the high-frequency cutoff of the low-frequency $1/f$ noise is given by the maximum relaxation rate of the TLSs, $\Gamma_{1,\rm max} \ll T$, as we assumed. Then the *extrapolations* of the low-frequency $1/f$ and high-frequency Ohmic spectra cross at this $\omega \sim T$.

We would like to emphasize that the relation between low- and high-frequency noise is more general, i.e., it is not unique to an ensemble of two-level systems. Consider an ensemble of many-level systems with levels $|n\rangle$ and energies E_n such that the coupling is via an observable which has both transverse and longitudinal components. By "transverse component" we mean the part constructed with operators $|n\rangle\langle m|$, where $n \neq m$, while the "longitudinal component" is built from the projectors $|n\rangle\langle n|$. If the system is under-damped, that is, if the absorption and emission lines are well defined, the correlator of such an observable will have Lorentzian-like contributions at $\omega = E_n - E_m$ as well as at $\omega = 0$. An example is provided by an ensemble of an-harmonic oscillators with $X = \sum_j v_j x_j$, where x_j are the oscillator's coordinates. Due to the anharmonicity x_j acquires a longitudinal component, in addition to the usual transverse one. Thus a relation between the low- and high-frequency noise would emerge naturally with details depending on the ensemble statistics.

4.6 Self-Consistent Model

In this section we consider a possibility that the Γ_1 decay of each individual TLS is caused by the other TLSs. This model explains further details of the behavior of $S_X(\omega)$. We assume that each individual fluctuator "feels" the same charge noise as

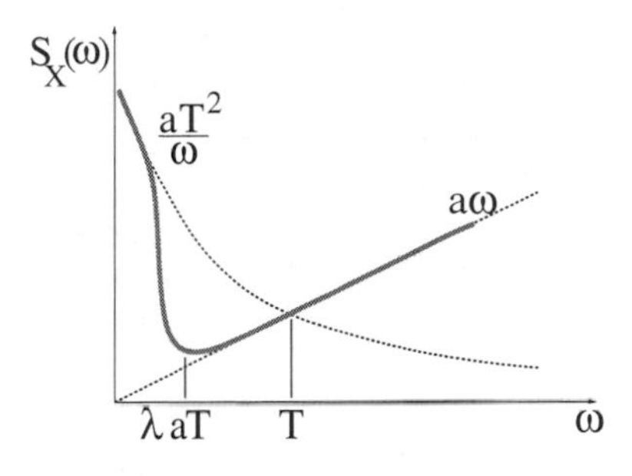

Fig. 4.3 Noise spectrum in a self-consistent model

the qubit, however reduced by a factor $\lambda < 1$ due to the small size of the fluctuators. That is we assume that the relaxation rate of the fluctuators is given by

$$\Gamma_{1,j} = \frac{\lambda}{2} \sin \theta_j^2 \, S_X(\omega = E_j). \tag{4.21}$$

As only the fluctuators with $E_j \leq T$ contribute to the 1/*f* noise, we estimate the maximum possible relaxation rate of the fluctuators to be $\Gamma_{1,\max} \sim \lambda a T$. This leads to a crossover from $1/f$ to $1/f^2$ dependence around $\omega \sim \Gamma_{1,\max} \sim \lambda a T$ as indicated in Fig. 4.3. We note that such a crossover (soft cut-off) is compatible with the recent experimental data [11].

4.7 Relation to Other Work

It is useful to relate our phenomenological results to the recent work of Faoro et al. [28], de Sousa et al. [29], Grishin et al. [30], and Faoro et al. [31], where physical models of the fluctuators, coupling to and relaxing the qubit, were considered. In Ref. [28] three models were studied: (I) a single electron trap in tunnel contact with a metallic gate, (II) a single electron occupying a double trap, and (III) a double trap that can absorb/emit a Cooper pair from the qubit or a superconducting gate ("Andreev fluctuator"). In all models a uniform distribution of the trap energy levels was assumed. One, then, can show that the distribution for the two-level systems corresponding to the models II and III are linear in the energy level splitting, $P(\epsilon) \propto \epsilon$. Since the switching in these models is tunneling dominated, we find that $P(\Delta) \propto 1/\Delta$. Therefore, both models II and III are characterized by distribution $P(\epsilon, \Delta) \propto \epsilon/\Delta$, introduced above, and hence can naturally account for the experimentally observed low- and high-frequency noises. In contrast, as shown in [29, 30], single electron traps do not behave as coherent two-level systems. Depending on the ratio between the hybridization with the metal and the temperature a single trap at the Fermi energy can either show the random telegraph noise or, when the hybridization dominates, it makes the qubit to feel the Ohmic particle-hole spectrum of the metal.

It was argued recently [31] that one needs an unphysically high density of fluctuators in order to explain the experimental findings. This argument is based on an assumption that the traps' energies are distributed homogeneously over the energy band of order of the Fermi energy (of order 1 eV). Faoro and Ioffe [31] proposed an alternative scenario where the low energy scale needed for qubit relaxation is provided by Kondo physics.

4.8 Conclusions

Josephson qubits have found their first application as sensitive meters of their environment. Measurements of qubit relaxation produced new surprising information about the properties of $1/f$ noise. Motivated by these experiments, we have shown that an ensemble of coherent two-level systems with the distribution function, $P(\epsilon, \Delta) \propto \epsilon/\Delta$, produces Ohmic high-frequency noise and, at the same time, $1/f$ low-frequency noise with strength which scales with temperature as T^2. The two branches of the noise power spectrum cross at $\omega \sim T$ in accordance with the experimental observation [10].

References

1. Collin, E., Ithier, G., Aassime, A., Joyez, P., Vion, D., Esteve, D.: Phys. Rev. Lett. **93**, 157005 (2004)
2. Astafiev, O., Pashkin, Yu.A., Yamamoto, T., Nakamura, Y., Tsai, J.S.: Phys. Rev. B **69**, 180507 (2004)
3. Siddiqi, I., Vijay, R., Pierre, F., Wilson, C.M., Metcalfe, M., Rigetti, C., Frunzio, L., Devoret, M.H.: Phys. Rev. Lett. **93**, 207002 (2004)
4. Yamamoto, T., Pashkin, Yu.A., Astafiev, O., Nakamura, Y., Tsai, J.S.: Nature 425, **941** (2003)
5. Wallraff, A., Schuster, D.I., Blais, A., Frunzio, L., Huang, R.-S., Majer, J., Kumar, S., Girvin, S.M., Schoelkopf, R.J.: Nature **431**, 162 (2004)
6. Esteve, D., Vion, D.: Solid state quantum bits, cond-mat/0505676 (2005)
7. Wendin, G., Shumeiko, V.S.: Superconducting quantum circuits, qubits and computing, cond-mat/0508729 (2005)
8. Aguado, R., Kouwenhoven, L.P.: Phys. Rev. Lett. **84**, 1986 (2000)
9. Schoelkopf, R.J., Clerk, A.A., Girvin, S.M., Lehnert, K.W., Devoret, M.H.: Qubits as spectrometers of quantum noise, cond-mat/0210247. In: Nazarov, Yu.V. (ed.) Quantum Noise in Mesoscopic Physics, pp. 175–203. Kluwer Academic Publishers, Dordrecht, Boston (2003)
10. Astafiev, O., Pashkin, Yu.A., Nakamura, Y., Yamamoto, T., Tsai, J.S.: Phys. Rev. Lett. **93**, 267007 (2004)
11. Ithier, G., Collin, E., Joyez, P., Meeson, P.J., Vion, D., Esteve, D., Chiarello, F., Shnirman, A., Makhlin, Yu., Schriefl, J., Schön, G.: Phys. Rev. B **72**, 134519 (2005)
12. Martinis, J.M., Cooper, K.B., McDermott, R., Steffen, M., Ansmann, M., Osborn, K., Cicak, K., Oh, S., Pappas, D.P., Simmonds, R.W., Yu, C.C.: Phys. Rev. Lett. **95**, 210053 (2005)
13. Bloch, F.: Phys. Rev. **105**, 1206 (1957)

14. Redfield, A.G.: IBM J. Res. Dev. **1**, 19 (1957)
15. Cottet, A.: PhD thesis, Université Paris VI (2002)
16. Shnirman, A., Makhlin, Yu., Schön, G.: Physica. Scripta **T102**, 147 (2002)
17. Paladino, E., Faoro, L., Falci, G., Fazio, R.: Phys. Rev. Lett. **88**, 228304 (2002)
18. Galperin, Y.M., Altshuler, B.L., Shantsev, D.V.: Low-frequency noise as a source of dephasing of a qubit. In: Lerner, I.V., Altshuler, B.L., Gefen, Yu. (eds.) Fundamental Problems of Mesoscopic Physics. Kluwer Academic Publishers, Dordrecht, Boston, London, cond-mat/0312490 (2004)
19. Wellstood, F.C.: PhD thesis. University of California, Berkeley (1988)
20. Kenyon, M., Lobb, C.J., Wellstood, F.C.: J. Appl. Phys. **88**, 6536 (2000)
21. Astafiev, O., et al.: Private communication (2004)
22. Wellstood, F.C., Urbina, C., Clarke, J.: Appl. Phys. Lett. **85**, 5296 (2004)
23. Simmonds, R.W., Lang, K.M., Hite, D.A., Nam, S., Pappas, D.P., Martinis, J.M.: Phys. Rev. Lett. **93**, 077003 (2004)
24. Martin, I., Bulaevskii, L., Shnirman, A.: Phys. Rev. Lett. **95**, 127002 (2005)
25. Shnirman, A., Schön, G., Martin, I., Makhlin, Yu.: Phys. Rev. Lett. **94**, 127002 (2005)
26. Dutta, P., Horn, P.M.: Rev. Mod. Phys. **53**, 497 (1981)
27. Phillips, W.A.: J. Low. Temp. Phys. **7**, 351 (1972)
28. Faoro, L., Bergli, J., Altshuler, B.A., Galperin, Y.M.: Phys. Rev. Lett. **95**, 046805 (2005)
29. de Sousa, R., Whaley, K.B., Wilhelm, F.K., Delft, J.v.: Phys. Rev. Lett. **95**, 247006 (2005)
30. Grishin, A., Yurkevich, I.V., Lerner, I.V.: Phys. Rev. B **72**, 060509 (2005)
31. Faoro, L., Ioffe, L.B.: Phys. Rev. Lett. **96**, 047001 (2006)

Chapter 5
Scanning Tunneling Spectroscopy

Markus Morgenstern

5.1 Introduction

Since the discovery of the scanning tunneling microscope [1], a huge number of different solid state samples have been investigated in order to understand the properties of the systems down to the atomic scale. Basically, a metallic tip is brought close to a sample surface, i.e., the distance between tip and sample is about 5 Å. At this distance a stable tunneling current between the tip and the sample is established, which couples the electronic states of the tip to the electronic states of the sample. Scanning the tip across the surface, thus, allows to study the local electronic properties at the sample surface, i.e., one gets insight into the local distribution of the electronic states. By using lock-in technique, one can separate the electronic states at different energies [2]. Provided that the tip injects electrons of a preferential spin orientation, one even gets additional information on the spin orientation of the corresponding electronic states [3].

5.2 Measuring the Local Density of States

The simplest description of the tunneling current is given by the Tersoff–Haman model [4, 5]. It requires that tip and sample can be considered as independent systems, which is a good approximation at distances above 4 Å. Moreover, it neglects any inelastic tunneling processes, i.e., tunneling is described in first-order perturbation. Finally, it assumes that the electronic states of the tip are spherically symmetric. Under these conditions, the tunneling current I is proportional to the

M. Morgenstern (✉)
II. Institute of Physics B, RWTH Aachen, Huyskensweg, 52056 Aachen, Germany
e-mail: mmorgens@physik.rwth-aachen.de

M. Vojta et al. (eds.), *CFN Lectures on Functional Nanostructures – Volume 2*,
Lecture Notes in Physics 820, DOI: 10.1007/978-3-642-14376-2_5

density of states of the sample, D_{sample}, measured at the symmetry point of the tip states, $\mathbf{r}_{\text{tip}}$, and integrated over the energy region between the Fermi level of the tip, $E_{\text{F,tip}}$, and the Fermi level of the sample, $E_{\text{F,sample}}$. At each energy, D_{sample} has to be multiplied by the density of states of the tip, D_{tip}, since tunneling is a joined process between electronic states of sample and tip.

$$I(\mathbf{r}_{\text{tip}}) \propto \int_{E_{\text{F,sample}}}^{E_{\text{F,tip}}} D_{\text{sample}}(E, \mathbf{r}_{\text{tip}}) \cdot D_{\text{tip}}(E) dE. \tag{5.1}$$

This equations holds at $T = 0$ K, but higher temperatures can easily be considered by using Fermi distribution functions within the integrand. The distance between the Fermi levels, $E_{\text{F,sample}} - E_{\text{F,tip}}$, is given by the applied voltage V multiplied by the electron charge e.

The property of interest is D_{sample}. Consequently, lock-in technique is used in order to measure the differential conductance dI/dV. At sufficiently low voltage, the derivative of the integral is dominated by the integrand. Additional terms, which arise due to the fact that the derivative is by V and not by E, can be neglected if $D_{\text{tip}}(E)$ is structureless within the considered energy region [6]. One gets:

$$dI/dV(\mathbf{r}_{\text{tip}}, V) \propto D_{\text{sample}}(E - E_{\text{F,sample}} = eV, \mathbf{r}_{\text{tip}}) \cdot D_{\text{tip}}(E = E_{\text{F,tip}}). \tag{5.2}$$

Since D_{tip} neither depends on V nor on $\mathbf{r}_{\text{tip}}$, dI/dV is a direct measure of the density of states of the sample at the position of the tip. In order to relate $D_{\text{sample}}(eV, \mathbf{r}_{\text{tip}})$ to the density of states at the sample surface $D_{\text{surface}}(eV, x, y)$, one can use the fact that the density of states decays exponentially into vacuum:

$$D_{\text{sample}}(eV, \mathbf{r}_{\text{tip}}) = D_{\text{surface}}(eV, x, y) \cdot e^{-\alpha z}. \tag{5.3}$$

With this formula, a direct access to the density of states at the surface is provided. However, one has to keep in mind that the decay constant α can depend on energy E and position (x, y). Thus, one measures the so called local density of states $D_{\text{surface}}(eV, x, y)$ at the energy corresponding to the applied voltage V, if α does not depend on (x, y). The latter can be checked by measuring the spatial dependence of $I(z) \propto e^{-\alpha \cdot z}$. It turns out that homogeneous samples usually exhibit a spatially constant α except at the atomic scale [6].

Next, the assumptions of the Tersoff–Haman model have to be considered. As mentioned previously, the independence of the two electron systems, sample and tip, is a good approximation at large enough tip–sample distance. Inelastic tunneling processes occur [7], but they are of minor importance since they are of second order. The assumption of spherically symmetric tip states is more questionable, since the tip is usually a d-metal including p- and d-states. These states can contribute to the tunneling process, if they are directed towards the sample (p_z- or d_{z^2}-states). Chen has shown that a simple derivative rule applies [8–10]:

$$p_z - \text{state} \rightarrow dI/dV(\mathbf{r}_{\text{tip}}, V) \propto \frac{dD_{\text{sample}}(eV, \mathbf{r}_{\text{tip}})}{dz} \cdot D_{\text{tip}}(E_{\text{F,tip}}), \tag{5.4}$$

$$d_{z^2} - \text{state} \rightarrow dI/dV(\mathbf{r}_{\text{tip}}, V) \propto \frac{d^2 D_{\text{sample}}(eV, \mathbf{r}_{\text{tip}})}{dz^2} \cdot D_{\text{tip}}(E_{\text{F,tip}}). \tag{5.5}$$

The derivative along z results in an additional prefactor α in (3). Consequently, the spatial distribution of $dI/dV(x, y)$ depends only on $D_{\text{surface}}(x, y)$, if α is spatially constant. We conclude that the spatial dependence of $D_{\text{surface}}(x, y)$ can be measured, if α is spatially constant. Measuring the energy dependence of $D_{\text{surface}}(E)$ requires that α is energetically constant and, moreover, that $D_{\text{tip}}(E)$ is energetically constant. Neither is usually the case. However, α mostly depends only weakly on energy, if one measures at low voltage V with respect to the work function of tip and sample (4–5 eV). The error can be estimated to be lower than 10% at voltages below $V \simeq 200$ mV [6]. At higher voltage, an empirical formula is used to eliminate the influence of α [11]:

$$D_{\text{surface}}(E - E_{\text{F,sample}} = eV) \propto \frac{dI/dV(V)}{I(V)/V}. \tag{5.6}$$

The influence of $D_{\text{tip}}(E)$ is not controllable, since each microtip has a different density of states. Thus, one has to repeat the measurements with different tips and assumes that peaks, which are reproducably measured with different tips, arise from electronic states of the sample.

Remarkably, the measured quantity $D_{\text{surface}}(E - E_{\text{F,sample}} = eV, x, y)$ is directly related to the electronic wave functions at the sample surface, i.e., one measures the squared wave functions at the selected energy:

$$D_{\text{surface}}(E, x, y) \propto \sum |\Psi(E, x, y)_{\text{surface}}|^2. \tag{5.7}$$

Experimentally, the finite temperature has to be considered as well as the fact that dI/dV is usually measured by lock-in-technique with the help of an applied ac-voltage of strength V_{eff}. This results in a finite energy resolution. Approximately, the energy resolution function is a Gaussian function with full width at half maximum $\Delta E = \sqrt{(3.3 \cdot kT)^2 + (2.5 \cdot V_{\text{eff}})^2}$ [12, 13]. Using $E_{\text{F,sample}} = 0$ meV, this results in

$$dI/dV(V, x, y) \propto \sum_i |\Psi(E_i, x, y)_{\text{surface}}|^2 \cdot e^{2(eV - E_i)^2/\Delta E^2}. \tag{5.8}$$

The energy resolution ΔE can be measured using an isotropic superconductor such as Nb as one of the electrodes. Figure 5.1 shows an example using a Nb tip on a W surface. The measurement is performed at $T = 315$ mK and $V_{\text{eff}} = 20$ µV [12]. The BCS-gap of the superconductor is visible. The sharp peaks at the edge of the gap are extremely sensitive to ΔE. The solid line in Fig. 5.1 shows a fit using an energy resolution $\Delta E = 100$ µeV.

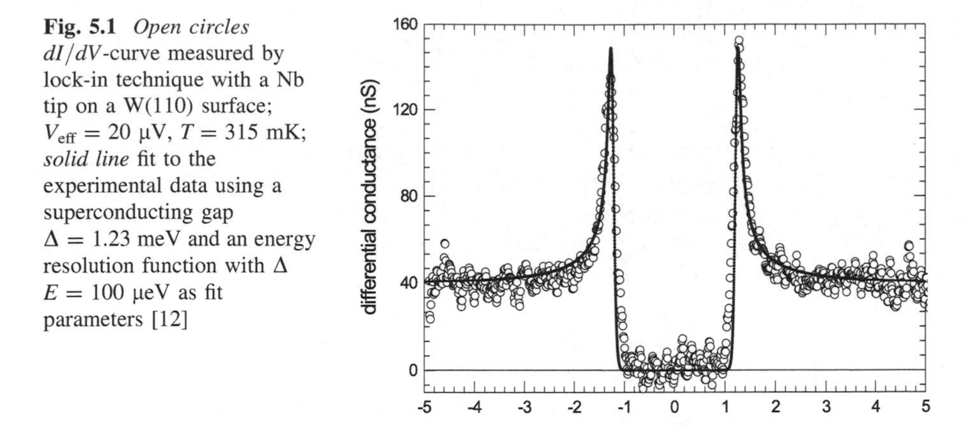

Fig. 5.1 *Open circles* dI/dV-curve measured by lock-in technique with a Nb tip on a W(110) surface; $V_{\text{eff}} = 20\ \mu\text{V}$, $T = 315$ mK; *solid line* fit to the experimental data using a superconducting gap $\Delta = 1.23$ meV and an energy resolution function with $\Delta E = 100\ \mu\text{eV}$ as fit parameters [12]

We conclude: as long as the decay constant α of the wave functions is spatially homogeneous, $dI/dV(V, x, y)$ measures the sum of all squared wave functions at the sample surface, which are in the energy region $E - E_{\text{F,sample}} = eV \pm \Delta E$.

Finally, it has to be mentioned that the above derivation requires that the tip–sample distance z is spatially constant. This is usually not the case, but one can use simple normalization processes [6, 14] to reconstruct the resulting $dI/dV(x, y, V)$-values at constant z.

The measurement of the local density of states extracted from the site- and energy-dependent differential conductance dI/dV is called scanning tunneling spectroscopy (STS).

5.3 Scanning Tunneling Spectroscopy on Semiconductors

As an example, STS measurements on InAs(110) are discussed. The wave functions in the InAs conduction band are relatively simple. The conduction band is nearly parabolic and isotropic, and the symmetry of the atomic wave functions is s-like [15]. The single-particle wave functions $\Psi(\mathbf{x}, t)$ can be described as Bloch waves:

$$\Psi(\mathbf{x}, t) = u_s(\mathbf{x}) \cdot e^{i(\mathbf{k}\cdot\mathbf{x} - \omega t)} \tag{5.9}$$

with

$$E = \frac{(\hbar\mathbf{k})^2}{2m^\star}. \tag{5.10}$$

Here, $u_s(\mathbf{x})$ is the atomically periodic part of the Bloch wave, $\mathbf{k}$ is the wave vector and $m^\star$ is the effective mass of the InAs conduction band ($m^\star \simeq 0.023 \times 9.1 \times 10^{-31}$ kg). The atomically periodic part $u_s(\mathbf{x})^2$ can be directly seen in dI/dV-images as shown in

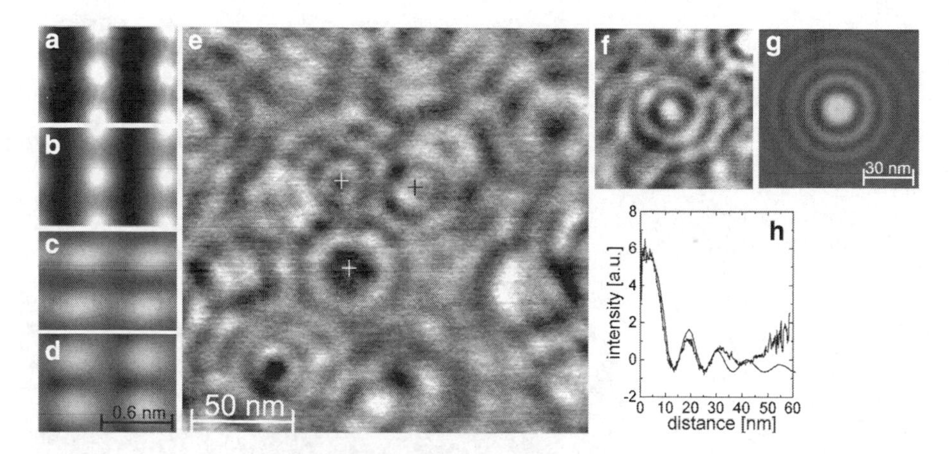

Fig. 5.2 **a** dI/dV-image of InAs(110) measured at $V = 100$ mV; **b** calculated dI/dV-image corresponding to **a**; **c** dI/dV-image of InAs(110) measured at $V = 900$ mV; **d** calculated dI/dV-image corresponding to **b**; **e** dI/dV-image measured at $V = 50$ mV; crosses mark positions of dopants; **f** dI/dV image measured at $V = 50$ mV showing the wave pattern around a defect located 14.3 nm below the surface; **g** calculated wave pattern corresponding to **f**; **h** height profiles of the measured and the calculated pattern; the x-axis starts from the center of the circular pattern (circular line section [14]); the measured pattern has been normalized in order to establish constant tip-surface distance; *scale bar* in **d** belongs to **a–d**, scale bar in **g** belongs to **f**, **g** [14, 15]

Fig. 5.2a, c. The function $u_s(\mathbf{x})^2$ can be calculated by density functional theory within the local density approximation. The result is shown in Fig. 5.2b, d. Obviously, the calculated patterns show good agreement with the measured ones [15]. The wave functions in Fig. 5.2a, b belong to the InAs conduction band and the wave functions in Fig. 5.2c, d belong to a surface band higher in energy [15].

The long range part $e^{i\mathbf{k}\cdot\mathbf{x}}$ can only be seen if the phase of the plane wave is fixed [16]. This can be realized by introducing defects into the atomically periodic structure of the crystal. One can argue that the plane wave is scattered at the defect, and the interference between the incoming and the reflected wave results in a standing electron wave. In semiconductors, the natural defects are dopants. Figure 5.2e shows a dI/dV image of InAs(110), which exhibits circular standing waves around the sulphur dopants. The circular structure is a direct consequence of the isotropic band structure of the InAs conduction band. The image displays the sum of all scattered waves at the corresponding energy. These are all wave functions with the same length of $|\mathbf{k}|$, but with $\mathbf{k}$ vectors pointing in different directions. Since one measures only the surface part of the scattered wave functions, different diameters of the circular structures result, which depend on the depth of the scatterer below the surface. A detailed analysis reveals that the measured patterns result from scatterers down to 25 nm below the surface [14]. The individual circular structures can be reproduced by a simple scattering theory as shown in Fig. 5.2f, h. Note that the measured wave functions belong to a bulk band of InAs. This is only possible if surface bands are absent in the investigated energy region. Otherwise, the surface states clearly dominate the measured pattern.

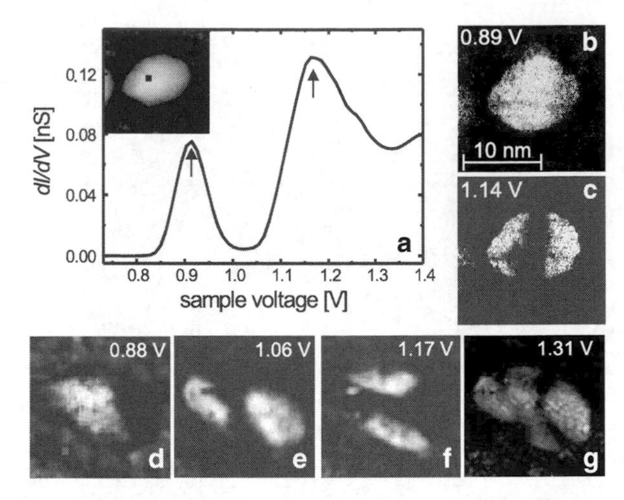

Fig. 5.3 **a** *dI*/*dV* curve measured on an InAs quantum dot; *arrows* mark the voltages where the *dI*/*dV* maps in **b** and **c** are taken; *inset* STM image of the quantum dot; *black dot* marks the position of the spectrum in the main image; **b**, **c** *dI*/*dV* maps taken at the voltages indicated and showing s- and p_x-like symmetry of the corresponding wave functions; **d**–**g** *dI*/*dV* maps taken on a different InAs quantum dot and showing s-,p_x-, p_y- and *d*-like symmetry; $T = 6$ K [18]

Within the previous example, each image consists of several single-particle wave functions. Individual single-particle wave functions can only be measured in a confined environment, e.g., in quantum dots. The requirement is, that the level spacing between the confined states is larger than the energy resolution of the experiment ΔE. For III–V semiconductors, this can be realized with strain-induced InAs quantum dots deposited on GaAs(001) [17]. Figure 5.3 shows an example [18]. The dI/dV curve in Fig. 5.3a is measured on a particular position of the quantum dot and shows two peaks belonging to two different single particle energies within the dot. The peaks are well separated indicating that only a single state $\Psi_i(x, y)$ contributes to $dI/dV(x, y)$ at a particular voltage. The $dI/dV(x, y)$ map recorded at the peak voltage represents the corresponding $|\Psi_i(x, y)|^2$, which is shown in Fig. 5.3b, c for the two peak voltages of Fig. 5.3a. The states obviously exhibit s- and p_x-like symmetry, respectively. Figure 5.3d–g show the squared wave functions of another InAs quantum dot exhibiting s-, p_x-, p_y- and d-like (2 nodes in *x*-direction) symmetry. More quantum dots have been studied, and it turns out that the shape asymmetry of the quantum dots has an important influence on the wave functions [18–20].

The wave functions of the InAs conduction band can also be used to study the influence of electronic interactions within an extended electron system. It is known that the interactions depend critically on the dimension of the system and on the magnetic field. The major problem is to provide the respective samples exhibiting 0D, 1D and 2D confinement. Additionally, the electron systems have to be close to the surface in order to get a tunneling current between the system and the STM tip. One possibility is to use naturally formed systems of lower dimension as they exist, e.g., on the InAs(110) surface. In order to get a two-dimensional electron system (2DES), one deposits minute amounts of adsorbates on the surface, which establish a 2D accumulation layer close to the surface [21]. The one-dimensional system (1DES) exists below [112]-step edges, which occur in distances of about 5 μm on the surface [22]. These step edges are positively charged and accumulate electrons in 1D. A zero-dimensional system (0DES) can be induced by using the tip as a gate [23]. One can achieve considerable work function differences between sample and tip. If the resulting tip charge is attractive for electrons, the tip accumulates electrons in the area around the tip, which are zero-dimensionally confined. Finally, a magnetic field can be applied externally by a superconducting coil.

The other parameters guiding the behavior of the electron system are the electron density and the strength of the electrostatic disorder. They are more difficult to tune within a STS experiment but can be determined experimentally [24]. Similar strengths of disorder have been achieved in different dimensions. A good measure of the disorder strength is given by the mean square deviation of the spatially fluctuating potential $U(x, y)$. The disorder strength is about 5 meV in the present experiments in 3D, 2D, 1D and 0D. The electron density is also similar in different dimensions, i.e., the electrons are on average 5–50 nm apart. From this distance, one can estimate that the electron–electron interaction energy is comparable to the disorder strength. Consequently, the electron system is guided by the interplay between optimizing disorder energy and electron–electron repulsion.

Measurements taken on a 1D system [22] are displayed in Fig. 5.4. An STM image of a [112] step edge is shown in (a), and the same area after subtracting the topographic height of the step is shown in (b). Bright and dark spots appear in (b), which indicate positively and negatively charged defects. This can be used to extract the effective electrostatic disorder potential of the 1D system [22], which is presented in Fig. 5.4c. The potential fluctuates between +40 and −40 meV with a mean square deviation of 10 meV.

The resulting local density of states $D_{\text{surface}}(\text{x, y, E}) \propto dI/dV\ (x, y, V)$ is shown in Fig. 5.4d–o for different energies. The images exhibit one-dimensional standing waves with decreasing wave length at higher energy. Since the black areas correspond to $D_{\text{surface}} \simeq 0$, we can conclude that the phases of nearly all wave functions are pinned by the disorder. Otherwise, the minima would have an intensity larger than zero. The fact that all wave functions are pinned is not surprising, because it is well known that wave functions in 1D are weakly localized, the terminus technicus for being standing waves.

As described above, each dI/dV image in Fig. 5.4 consists of several single-particle wave functions. It is not clear, whether these standing wave functions spill out of the image area, i.e., it might be that only part of the weight of an individual wave function is measured. On the other hand, one can easily calculate the total weight of wave functions within the image area, which turns out to correspond to about 2.5 electrons in each image [22].

Basically, the dI/dV images of the 1DES can be reproduced by a single-particle calculation. This is disappointing, since one of the most interesting effects in 1D systems appears if the electron–electron interaction guides the system. Under these conditions, the single-particle description breaks down leading to the so-called Luttinger properties [25]. But indications of these intriguing properties have not been found in this system probably due to the relatively strong disorder strength (10 meV) with respect to the electron–electron interaction energy (20 meV).

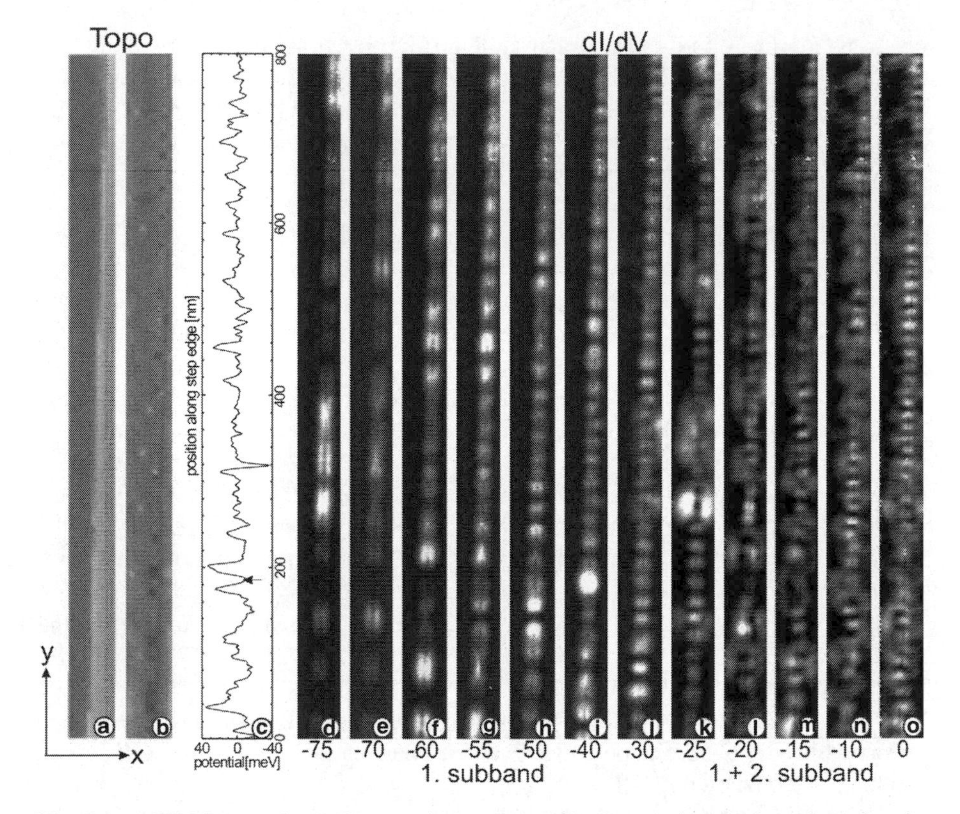

Fig. 5.4 **a** STM image of a [112] step edge on InAs(110); **b** same image after subtracting the average apparent height of the step edge; dopants are visible as bright and dark spots; **c** deduced disorder potential along the step edge; **d–o** dI/dV maps measured at the voltages marked in meV below the images; 1D standing waves indicating weakly localized states are visible; $T = 6$ K [22]

2D electron systems (2DES) can be either prepared by the growth of InAs on GaAs(111) [26] or by depositing minute amounts of adsorbates on InAs(110) [21]. In the former case, the interface to the GaAs realizes the 2D confinement, while in the latter case a surface accumulation layer is formed. Both systems provide relatively large electron densities ($\simeq 10^{12}/\text{cm}^2$) and a moderate disorder strength (5–20 meV) [27]. However, the influence of the interaction of the electrons with the disorder, which is known to lead to weak localization, can be measured on the local scale. Figure 5.5a shows a dI/dV image of a 2DES [27]. A complex wave pattern with a preferential wave length is visible. The Fourier transform of the real space data is shown in the right inset. It is a representation of the wave vectors contributing to the dI/dV image. The apparent ring structure demonstrates that the majority of the contributing wave vectors has the same length.

Figure 5.5c shows a histogram of all dI/dV values obtained in Fig. 5.5a. The histogram is broad, i.e., one finds many different $D_{\text{surface}}(x, y, E)$ values within the investigated sample area. $D_{\text{surface}}(x, y, E)$ spatially fluctuates between large values and small values, respectivly, the $D_{\text{surface}}(x, y, E)$ corrugation is large. Again, a

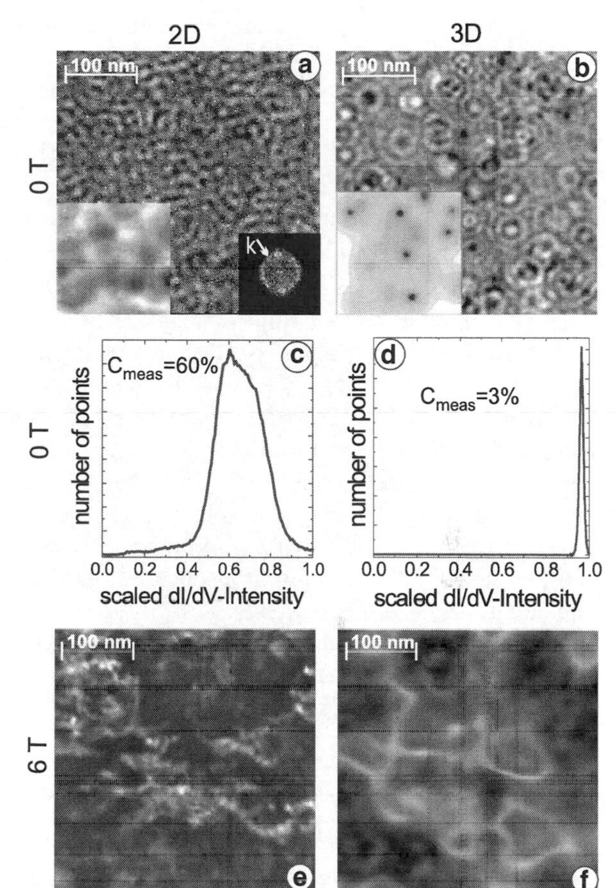

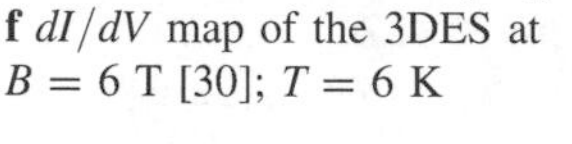

Fig. 5.5 **a** dI/dV map of a 2DES at $B = 0$ T; *insets*: *left* measured disorder potential of the 2DES; *right* Fourier transform of the LDOS with dominating **k**-vector length indicated [27]; **b** dI/dV map of a 3DES measured at the same kinetic energy as the 2DES at $B = 0$ T; *inset*: measured disorder potential of the 3DES at the surface [14]; **c**, **d** histograms of the data of **a**, **b** with indicated values of relative $D_{\mathrm{surface}}(x, y, E)$ corrugation (C_{meas} [13, 24]); **e** dI/dV map of the 2DES at $B = 6$ T [29]; **f** dI/dV map of the 3DES at $B = 6$ T [30]; $T = 6$ K

single particle calculation using the measured disorder potential (left inset in Fig. 5.5a) can reproduce the features of the dI/dV-image [27].

Since the pristine 3D InAs(110) system has also been measured [14], a direct comparison between 2D and 3D behaviour of electrons is possible. A dI/dV image of the 3D system and its histogram are shown in Fig. 5.5b, d. Note that both systems, 2D and 3D, are measured at the same temperature, with the same disorder strength and at the same energy with respect to the band edge. Thus, the comparison is rather direct. Obviously, the 3D pattern is much more regular, consisting of self-interference rings around individual scatterers. Moreover, the corrugation of the 3D system is an order of magnitude less. This is a direct visualization of the fact that 2D systems tend to weakly localize, while 3D systems do not weakly localize [28]. In 3D, the pattern results only from scattering processes at individual dopants, while in 2D more complex scattering paths including several dopants lead to the observed complex pattern. The fact that the scattering paths are closed in 2D [28] leads to the complete phase fixing of all electronic states resulting in the observed strong corrugation, and it induces the well-known weak localization in 2D.

As an important thermodynamic parameter, a magnetic field up to 6 T has been applied to systems from 0D to 3D. Landau and spin quantization have been observed for 2D and 0D systems [31, 32]. More interestingly, the transformation of the wave functions in a magnetic field could be observed in real space. While 0D and 1D systems did not show pronounced changes due to the relatively large confinement, a distinct change has been observed for 2D [29] and 3D [30]. In both cases, serpentine structures exhibiting strong corrugation appear as shown in Fig. 5.5e, f. The full width at half maximum of the serpentine widths is exactly the magnetic length $l_B = \sqrt{\hbar/(eB)}$. Moreover, in 2D, the patterns are periodic in energy having a periodicity of the Landau energy. Thus, the theoretically predicted drift states [33, 34], which arise due to the interaction of Lorentz forces and electrostatic disorder, are experimentally confirmed. The states run along equipotential lines of the disorder potential as predicted [35]. The fact that drift states appear also in 3D, was surprising and could be linked to the appearance of a Coulomb gap at the Fermi level [36]. Probably, a partial localization of the electrons parallel to the magnetic field arises in the extreme quantum limit prior to magnetic freeze out. This could lead to local 2D properties of the electrons [37]. Thus, electrons running along equipotential lines of the disorder also exist in 3D at sufficiently strong magnetic fields.

The presented results demonstrate the possibilities of STS to directly image the wave functions in interacting electron systems. Here, we used III–V-semiconductors as an example, but other systems including such complex ones as Kondo systems [38–40] or high-temperature superconductors [41–43] have been successfully investigated on the local scale. A summary of recent results is given in [44].

5.4 Spin-Polarized Scanning Tunneling Spectroscopy

In the previous section, we described STS measurements which record the sum of squared wave functions $\Psi_i(E_i, x, y)_{\text{surface}}$ at the selected energy $E = eV$. The squared wave functions describe the charge distribution of the electronic states. However, the spin of the electron might also be important for the behavior of the electron system, respectively, might be used for information processing [45]. Consequently, it is desirable to measure the spin distribution of the electrons as well.

A suitable extension of STS has been proposed by Wiesendanger et al. [46, 47] and has been optimized recently [3]. Instead of using a conventional metallic tip, one covers the tip with a thin layer of ferromagnetic material acting as a spin filter. This leads to a difference of $D_{\text{tip}}(E)$ for spin-down and spin-up electrons: $D_{\text{tip},\downarrow}(E) \neq D_{\text{tip},\uparrow}(E)$. Since the tunneling process is spin conserving (at least in first order), (2) has to be replaced by:

$$dI/dV(\mathbf{r}_{\text{tip}}, V) \propto \sum_{i=\uparrow,\downarrow} D_{\text{sample},i}(E = eV, \mathbf{r}_{\text{tip}}) \cdot D_{\text{tip},i}(E = E_{\text{F,tip}}) \qquad (5.11)$$

The differential conductance dI/dV is high if the preferential spins of sample and tip are parallel, and low if they are antiparallel. This gives rise to a spatial $dI/dV(x, y)$ contrast between areas of parallel and antiparallel spin orientation. If the charge distribution of the electronic states is spatially constant as, e.g., in conventional ferromagnets, this contrast can be directly used to observe spin domains.

In order to get quantitative information, one has to perform two measurements with different spin orientations of the tip. For example, one can switch the preferential spin orientation of a ferromagnetic tip by applying an external B-field. Then, the difference between the dI/dV-values measured by the two spin configurations $\Delta dI/dV(V,x,y) = dI/dV_{\uparrow}(V,x,y) - dI/dV_{\downarrow}(V,x,y)$ corresponds to the spin density of the sample surface $S_{\text{surface}}(E,x,y) := D_{\text{surface},\uparrow}(E,x,y) - D_{\text{surface},\downarrow}(E,x,y)$ multiplied by the spin density of the tip $S_{\text{tip}}(E_{\text{F,tip}}) := D_{\text{tip},\uparrow}(E_{\text{F,tip}}) - D_{\text{tip},\downarrow}(E_{\text{F,tip}})$:

$$\Delta dI/dV(V,x,y) \propto S_{\text{surface}}(E,x,y) \cdot S_{\text{tip}}(E_{\text{F,tip}}). \tag{5.12}$$

Similar requirements as in Sect. 5.2 are necessary, i.e., a spatially constant α is required in order to measure the $S_{\text{surface}}(x, y)$ dependance, and an energetically constant α and S_{tip} are required in order to measure the $S_{\text{surface}}(E)$ dependance. Moreover, the decay constant α has to be the same for both spin orientations of the sample states. But even if α depends on the spin orientation, the measurement of $\Delta dI/dV$ can be used to disentangle spin information and charge information.

The measurement of $\Delta dI/dV$ has been implemented by positioning a coil around a ferromagnetic tip. An ac-current is applied to the coil in order to switch the spin orientation of the tip. The resulting $\Delta dI/dV$ is directly measured by lock-in technique [48].

Within these measurements, one has to take care of the fact that a ferromagnetic tip produces a magnetic stray field, which can influence the sample. However, as described in Sect. 5.2, the relevant $D_{\text{tip}}(E)$ is the one at the symmetry point of the tip $\mathbf{r}_{\text{tip}}$, which is usually the center of the atom closest to the sample surface. Thus, one can use antiferromagnetic tip coatings (Cr), which do not exhibit stray fields, but still provide a sufficient S_{tip} at the very last atom [49].

The above derivation implies that tip and sample provide the same spin axis. This might not be the case for, e.g., ferromagnetic domain structures, for which the spin axis spatially fluctuates. However, if the spin axis of the tip and the spin axis of the sample include an angle $\theta(x, y)$, (12) rewrites [3]:

$$\Delta dI/dV(V,x,y) \propto S_{\text{surface}}(E,x,y) \cdot S_{\text{tip}}(E_{\text{F,tip}}) \cdot \cos\theta(x,y). \tag{5.13}$$

In conventional ferromagnets $S_{\text{surface}}(x, y)$ is spatially constant resulting in $\Delta dI/dV(V,x,y) \propto \cos\theta(x,y)$. Thus, one measures the spatial distribution of the spin axis of the sample, which is called magnetization direction. Note that only the polar angle θ with respect to the spin axis of the tip is measured, i.e., the azimuthal angle ϕ of the magnetization direction cannot be determined. If the magnetization direction only rotates in the plane parallel to the surface, one needs a tip with a spin axis parallel to the surface plane in order to observe a spin contrast. If the sample spins rotate between the parallel and the perpendicular direction, one needs a tip with a perpendicular spin axis. These different types of tips can be prepared reproducably by using different amounts of coating material (Cr) on the tip [49].

The technique to measure the local spin density of states either by dI/dV or by $\Delta dI/dV$ is called spin-polarized scanning tunneling spectroscopy (SPSTS).

5.5 SPSTS on Ferromagnetic Nanostructures

As an example, the measurement of the magnetic domain configuration of Fe islands deposited on W(110) is discussed [50]. The islands exhibit heights of 3–10 nm and lateral dimensions of several 100 nm. Due to the interplay between stray field energy and magnetocrystalline anisotropy, different domain patterns result from different island shapes [51].

First, a tip with a spin axis parallel to the sample surface is used in order to be sensitive to the in-plane magnetization direction of the sample. Figure 5.6 shows the results. The first column displays STM measurements showing the topography of the islands. A line section across one island is given for each image in the third column revealing that the islands exhibit different heights. The dI/dV maps are visible in the second column. Here, it is sufficient to display dI/dV in order to show the spin contrast, since it has been checked by STS measurements using conventional W tips that the charge distribution of the density of states is homogeneous on each island.

The following results have been found. The islands in the first row exhibit only one shade of grey, i.e., the magnetization direction is constant on each island, as indicated in the fourth column. However, islands of greyscale value are observed corresponding to islands of spin direction parallel and antiparallel to the spin direction of the tip. We conclude that each island consists of a single magnetic domain. The islands in the second row are slightly higher, and the left one shows two areas of different shade of gray corresponding to two domains. This is again indicated in the fourth column. The island in the third row is even higher, and four areas are detectable. These are assigned to the four domains displayed in the fourth column. Note that the left and the right domain exhibit the same shade of gray, since they both have a polar angle $\theta = 90°$ with respect to the spin axis of the tip (left–right direction), but only a different azimuthal angle of $\phi = +90°$ and $\phi = -90°$, which is not detectable. Finally, the fourth row shows the highest island and the most complex pattern, probably consisting of seven domains as indicated in the fourth column. Here, one needs additional micromagnetic calculations to reveal the correct domain structure due to the fact that one is not sensitive to the azimuthal angle.

An interesting effect can be observed on the island in the third row. Here, all spins are oriented parallel to the edge of the island resulting in a zero stray field. Such a configuration is called a flux closure configuration. The energetic problem of the flux closure configuration is the center of the island. The spins have to rotate

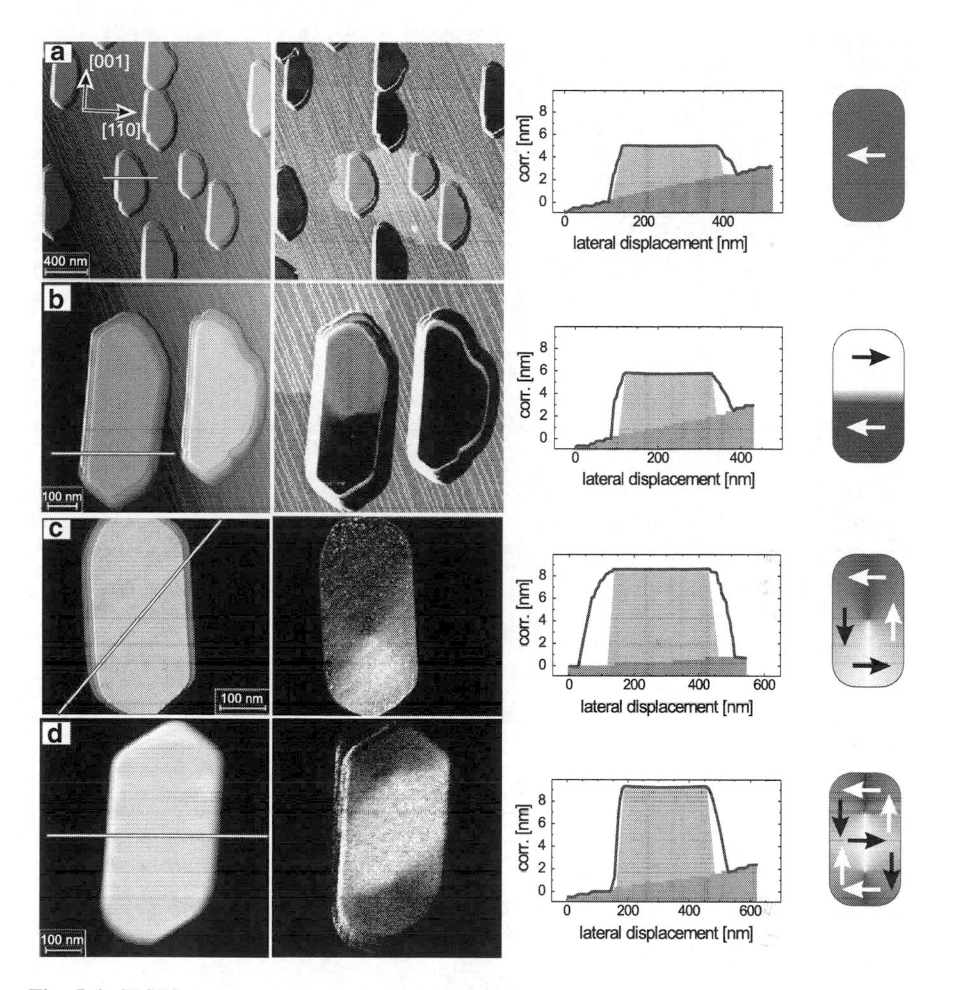

Fig. 5.6 SPSTS measurements on Fe islands deposited on W(110); *each row* shows a different height of the islands; *first column* STM measurements revealing the topography of the islands; *second column* dI/dV maps measured with a tip being sensitive to the in-plane orientation of magnetization; *third column* line sections measured along the lines in the first row; *fourth column* schematic sketches of the magnetic domain configurations of the islands [50]

continuously on a circle around the center as in a vortex. If the circle gets smaller towards the center, adjacent spins on the circle exhibit larger and larger relative angles. In a usual 2D vortex the angle would diverge in the center leading to a large exchange energy. It has been proposed that the exchange energy can be reduced by turning the spins in the center out-of-plane as sketched in Fig. 5.7a [52]. This is indeed observed by measuring the spin contrast in the central region twice, once with a tip being sensitive to the in-plane spin orientation of the sample (Fig. 5.7b) and once with a tip being sensitive to the out-of-plane orientation (Fig. 5.7c). In the out-of-plane configuration, only a bright spot of 9 nm in diameter is visible. This directly proves the theoretical prediction of spins turning

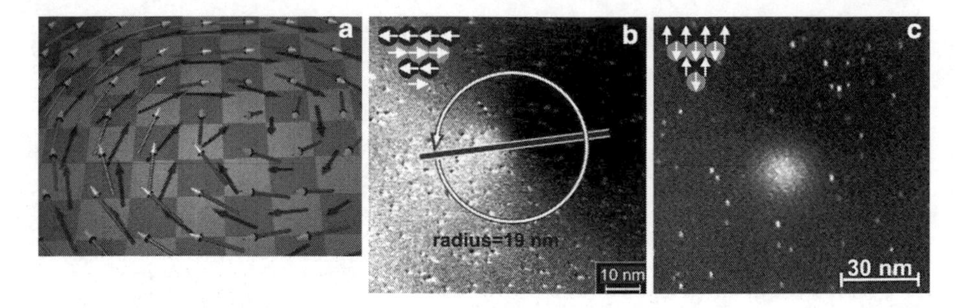

Fig. 5.7 SPSTS measurements in the center of an Fe island exhibiting a flux closure configuration: **a** sketch of the predicted spin configuration; **b** dI/dV map measured with a tip being sensitive to the in-plane orientation of the spins of the sample as sketched in the *inset*;**c** dI/dV map measured with a tip being sensitive to the out-of-plane orientation of the spins of the sample [52]

out-of-plane. Indeed, the diameter of the out-of-plane area and the measured distribution of $\theta(x, y)$ exactly reproduce the predictions from micromagnetic analysis [53, 54].

It has been shown that the spin contrast can be observed down to the atomic scale, e.g., by measuring the alternating spin orientation of adjacent atoms in antiferromagnets [55, 56]. Thus, SPSTS is currently the spin sensitive technique with the highest spatial resolution. Other techniques such as, e.g., Lorentz microscopy have a spatial resolution which is more than an order of magnitude worse.

5.6 Conclusions

The examples shown demonstrate that STS and SPSTS are powerful techniques to measure the spatial distribution of the sum of single-particle wave functions. The possible energy resolution is down to 100 μeV, and the spatial resolution is down to the atomic scale. STS is sensitive to the charge distribution of the wave functions, while SPSTS is sensitive to the spin distribution of the wave functions. Complex electronic structures can be measured provided they are located close enough to the sample surface.

Certainly, the combination of these two techniques will continue to give additional insight into the complex and fascinating quantum behavior of electron systems guided by different interaction energies. This requires ongoing effort to prepare the right samples in a fashion compatible with the UHV- and near-to-surface requirements of STS and SPSTS.

Acknowledgments I gratefully acknowledge that the presented data has been measured and calculated by several collaborators, in particular T. Maltezopolous, J. Klijn, C. Wittneven, R. Dombrowski, D. Haude, C. Meier, A. Wachowiak, J. Wiebe, and L. Sacharow. I appreciate useful discussions with M. Liebmann. Finally, I acknowledge the financial support by the Deutsche Forschungsgemeinschaft (SFB 508-B4 and A6, DFG Wi 1277/15).

References

1. Binnig, G., Rohrer, H., Gerber, Ch., Weibel, E.: Appl. Phys. Lett. **40**, 178 (1982)
2. Wiesendanger, R.: Scanning Probe Microscopy and Spectroscopy, 2nd edn. Cambridge University Press, Cambridge (1998)
3. Bode, M.: Rep. Prog. Phys. **66**, 523 (2003)
4. Tersoff, J., Hamann, D.R.: Phys. Rev. Lett. **50**, 1998 (1983)
5. Tersoff, J., Hamann, D.R.: Phys. Rev. B **31**, 805 (1985)
6. Morgenstern, M., Haude, D., Gudmundson, V., Wittneven, Chr., Dombrowski, R., Steinebach, Chr., Wiesendanger, R.: J. Electron Spectrosc. Relat. Phenom. **109**, 127 (2000)
7. Stipe, B.C., Rezaei, M.A., Ho, W.: Science **280**, 1732 (1998)
8. Chen, C.J.: Introduction to Scanning Tunneling Microscopy. Oxford University Press, Oxford (1993)
9. Chen C.J.: J. Vac. Sci. Technol. A **6**, 319 (1988)
10. Chen, C.J.: J. Vac. Sci. Technol. A **9**, 44 (1991)
11. Feenstra, R.M.: Phys. Rev. B **50**, 4561 (1994)
12. Wiebe, J., Wachowiak, A., Meier, F., Haude, D., Foster, T., Morgenstern, M., Wiesendanger, R.: Rev. Sci. Instrum. **75**, 4871 (2004)
13. Morgenstern, M.: Scanning tunneling spectroscopy. In: Vilarinho, P. (ed.) Scanning Probe Microscopy: Characterization, Nanofabrication and Device Application of Functional Materials, pp. 251–273. Kluwer, Amsterdam (2005)
14. Wittneven, Chr., Morgenstern, M., Dombrowski, R., Wiesendanger, R.: Phys. Rev. Lett. **81**, 5616 (1998)
15. Klijn, J., Sacharow, L., Meyer, Chr., Blügel, S., Morgenstern, M., Wiesendanger, R.: Phys. Rev. B **60**, 205327 (2003)
16. Crommie, M.F., Lutz, C.M., Eigler, D.M.: Nature **363**, 524 (1993)
17. Bimberg, D.: Quantum Dot Heterostructures. Wiley, New York (1999)
18. Maltezopoulos, Th., Bolz, A., Meyer, Chr., Heyn, Ch., Hansen, W., Morgenstern, M., Wiesendanger, R.: Phys. Rev. Lett. **91**, 196804 (2004)
19. Grandidier, B., Niquet, Y.M., Legrand, B., Nys, J.P., Priester, C., Stievenard, D.: Phys. Rev. Lett. **85**, 1068 (2000)
20. Millo, O., Katz, D., Cao, Y.W., Banin, U.: Phys. Rev. Lett. **86**, 5751 (2001)
21. Aristov, V.Yu., LeLay, G., Soukiassian, P., Hricovini, K., Bonnet, J.E., Osvald, J., Olsson, O.: Europhys. Lett. **26**, 359 (1994)
22. Meyer, Chr., Klijn, J., Morgenstern, M., Wiesendanger, R.: Phys. Rev. Lett. **91**, 76803 (2003)
23. Dombrowski, R., Steinebach, Chr., Wittneven, Chr., Morgenstern, M., Wiesendanger, R.: Phys. Rev. B **59**, 8043 (1999)
24. Morgenstern, M.: Surf. Rev. Lett. **10**, 933 (2003)
25. Luttinger, J.M.: J. Math. Phys. **4**, 1154 (1963)
26. Kanisawa, K., Butcher, M.J., Yamaguchi, H., Hirayama, Y.: Phys. Rev. Lett. **86**, 3384 (2001)
27. Morgenstern, M., Klijn, J., Meyer, Chr., Getzlaff, M., Adelung, R., Rossnagel, K., Kipp, L., Skibowski, M., Wiesendanger, R.: Phys. Rev. Lett. **89**, 136806 (2002)
28. Abrahams, E., Anderson, P.W., Licciardello, D.C., Ramakrishnan, T.V.: Phys. Rev. Lett. **42**, 673 (1979)
29. Morgenstern, M., Klijn, J., Meyer, Chr., Wiesendanger, R.: Phys. Rev. Lett. **90**, 56804 (2003)
30. Haude, D., Morgenstern, M., Meinel, I., Wiesendanger, R.: Phys. Rev. Lett. **86**, 1582 (2001)
31. Hashimoto, K., Wiebe, J., Meier, F., Morgenstern, M., Wiesendanger, R.: (to be published) (2010)
32. Morgenstern, M., Gudmundsson, V., Dombrowski, R., Wittneven, Chr., Wiesendanger, R.: Phys. Rev. B **63**, 201301 (2001)
33. Joynt, R., Prange, R.E.: Phys. Rev. B **29**, 3303 (1984)
34. Prange, R.E.: Effects of imperfections and disorder. In: Prange, R.E., Girvin, S.M. (eds.) The Quantum Hall Effect, pp. 69–98. Springer, New York (1987)

35. Ando, T.: J. Phys. Soc. Jpn. **53**, 3101 (1984)
36. Morgenstern, M., Haude, D., Klijn, J., Wiesendanger, R.: Phys. Rev. B **56**, 121102 (2002)
37. Morgenstern, M., Haude, D., Meyer, Chr., Wiesendanger, R.: Phys. Rev. B **64**, 205104 (2001)
38. Li, J., Schneider, W.D., Berndt, R., Delley, B.: Phys. Rev. Lett. **80**, 2893 (1998)
39. Madhavan, V., Chen, W., Jamneala, T., Crommie, M.F., Wingreen, N.S.: Science **280**, 567 (1998)
40. Manoharan, H.C., Lutz, C.P., Eigler, D.M.: Nature **403**, 512 (2000)
41. Pan, S.H., Hudson, E.W., Lang, K.M., Eisaki, H., Uchida, S., Davis, J.C.: Nature **403**, 746 (2000)
42. Lang, K.M., Madhavan, V., Hoffmann, J.E., Hudson, E.W., Eisaki, H., Uchida, S., Davis, J.C.: Nature **415**, 412 (2002)
43. Hoffmann, J.E., Hudson, E.W., Lang, K.M., Madhavan, V., Eisaki, H., Uchida, S., Davis, J.C.: Science **295**, 466 (2002)
44. Morgenstern, M., Schwarz, A., Schwarz, U.D.: Low temperature scanning probe microscopy. In: Bhushan, B. (ed.) Handbook of Nanotechnology, pp. 413–448. Springer, Berlin (2004)
45. Prinz, G.A.: Phys. Today **48**, 58 (1995)
46. Wiesendanger, R., Güntherodt, H.J., Güntherodt, G., Gambino, R.J., Ruf, R.: Phys. Rev. Lett. **65**, 247 (1990)
47. Wiesendanger, R., Shvets, I.V., Bürgler, D., Tarrach, G., Güntherodt, H.J., Coey, J.M.D., Gräser, S.: Science **255**, 583 (1992)
48. Wulfhekel, W., Kirschner, J.: Appl. Phys. Lett. **75**, 1944 (1999)
49. Kubetzka, A., Bode, M., Pietzsch, O., Wiesendanger, R.: Phys. Rev. Lett. **88**, 057201 (2002)
50. Bode, M., Wachowiak, A., Wiebe, J., Kubetzka, A., Morgenstern, M., Wiesendanger, R.: Appl. Phys. Lett. **84**, 948 (2004)
51. Hertel, R.: Z. Metallkd. **93**, 957 (2002)
52. Wachowiak, A., Wiebe, J., Bode, M., Pietzsch, O., Morgenstern, M., Wiesendanger, R.: Science **288**, 577 (2002)
53. Feldtkeller, E., Thomas, H.: Phys. Condens. Mater. **4**, 8 (1965)
54. Hubert, A., Schäfer, R.: Magnetic Domains. Springer, Berlin (1998)
55. Heinze, S., Bode, M., Kubetzka, A., Pietzsch, O., Nie, X., Blügel, S., Wiesendanger, R.: Science **288**, 1805 (2000)
56. Kubetzka, A., Ferriani, P., Bode, M., Heinze, S., Bihlmayer, G., Bergmann, K.v., Pietzsch, O., Blügel, S., Wiesendanger, R.: Phys. Rev. Lett. **94**, 87204 (2005)

Chapter 6
Manipulating Single Spins in Quantum Dots Coupled to Ferromagnetic Leads

Jürgen König, Matthias Braun and Jan Martinek

6.1 Introduction

The study of single spins in quantum-dot spin valves resides at the intersection of the two highly interesting and extensively pursued research fields of spintronics on the one hand and transport through nanostructures on the other hand. Quantum dots consist of a small confined island with a low capacitance such that a macroscopic gate or bias voltage is needed to add a single electron, leading to Coulomb-blockade phenomena [1–3]. The notion that not only the charge but, simultaneously, also the spin degree of freedom of the electrons can be made use of, for example by using ferromagnetic leads, defines the field of spintronics [4, 5]. A quantum-dot spin valve, i.e., a quantum dot coupled to ferromagnetic leads, exploits both the spin polarization of the electrons and the sensitivity of the charge

J. König (✉) and M. Braun
Institut für Theoretische Physik III, Ruhr-Universität Bochum, 44780 Bochum, Germany
e-mail: koenig@thp.uni-due.de

J. König
Fakultät für Physik, Universität Duisburg-Essen, Campus Duisburg,
47048 Duisburg, Germany

M. Braun
Areva S.A., Paris, France
e-mail: matthias.braun@areva.com

J. Martinek
Institute of Molecular Physics, Polish Academy of Science, 60-179 Poznań, Poland

J. Martinek
Institut für Theoretische Festkörperphysik, Karlsruhe Institut für Technologie (KIT),
76128 Karlsruhe, Germany

J. Martinek
Institute for Materials Research, Tohoku University, Sendai 980-8577, Japan

Vojta et al. (eds.), *CFN Lectures on Functional Nanostructures – Volume 2*,
Lecture Notes in Physics 820, DOI: 10.1007/978-3-642-14376-2_6,

to Coulomb interaction. Therefore, electronic transport is governed by the behavior of a *single spin*. To discuss the possibility to generate, manipulate, and probe single spins via electronic transport through quantum-dot spin valves is the goal of this chapter.

The capacitance C of a metallic or semiconductor island decreases when shrinking its size. For small quantum dots, the energy scale to add a single electron to the island, the charging energy $U = e^2/2C$, exceeds the energy scales set by temperature $k_B T$ or bias voltage eV, and Coulomb-blockade phenomena arise, as first observed by Fulton and Dolan [6]. If, in addition, the island size becomes comparable to the Fermi wavelength then the level spectrum on the island will be discrete. For sufficiently large energy-level spacings, only a single level may participate in transport. Such as system can then be described by the Anderson impurity model, which is introduced below.

Famous examples of spintronics devices are the spin valves based on either the giant magnetoresistance effect [7] in magnetic multilayers or the tunnel magnetoresistance [8] in magnetic tunnel junctions. These effects arise, when two ferromagnetic leads are in contact via a conducting layer or a tunnel barrier, respectively. The transport characteristics of the device then depend on the relative orientations of the lead magnetizations. If, in a magnetic tunnel junction, the lead magnetizations enclose the angle ϕ, the conductance through the tunnel junction is proportional to $\cos\phi$ [9–11], i.e., it is maximal for parallel and minimal for antiparallel alignment of the leads' magnetizations. This angular dependence simply reflects the overlap of the spinor part of the majority-spin wave functions in the source and drain electrode, which is given by the externally controlled leads' magnetizations.

This picture changes once spin accumulation can occur. Let us consider transport through a ferromagnet–nonmagnet–ferromagnet sandwich structure with the thickness of the normal layer being smaller than the spin diffusion length. A finite bias voltage applied between the two ferromagnets with nonparallel magnetization directions leads to a local imbalance of spin-up and spin-down electrons in the nonmagnetic layer. This non-equilibrium polarization of the electrons in the nonmagnetic region, known as spin accumulation, mediates the information of the relative orientation of the leads' magnetization through the middle part, such that the transmission through the device is reduced for increasing angle ϕ between directions of the leads' magnetic moments.

An extreme limit of the above scheme is realized in a quantum-dot spin valve. It consists of a quantum dot that is tunnel coupled to ferromagnetic leads, see Fig. 6.1. In this case, the information about the leads' relative magnetization directions is mediated by a single quantum-dot electron that, as a consequence of a finite bias voltage, is partially spin polarized, described by a finite quantum-statistical average $\mathbf{S} = (\hbar/2)\langle\boldsymbol{\sigma}\rangle$ of the dot spin. It is the orientation of the dot spin relative to the leads together with the degree of the dot spin polarization that determines the transport, rather than just the relative orientation of the leads'

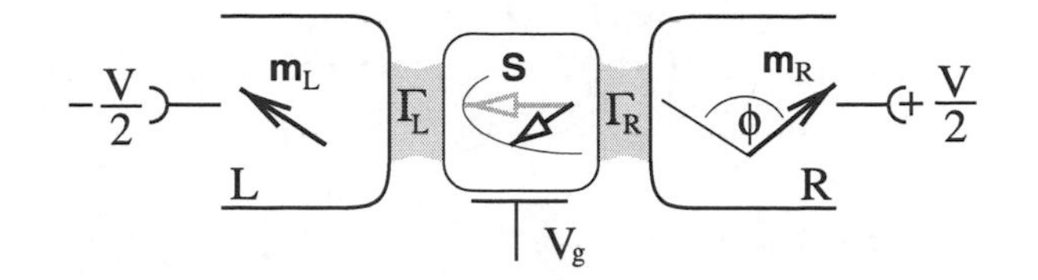

Fig. 6.1 A quantum dot contacted by ferromagnetic leads with non-collinear magnetizations. The lead magnetization directions enclose the angle ϕ. By forcing a current through the system, a non-equilibrium spin **S** accumulates on the otherwise non-magnetic dot

magnetization directions only. Any manipulation of the dot spin polarization will change the transmission through the device.

A quantum-dot spin valve is, thus, a convenient tool to generate, manipulate, and detect spin polarization of single quantum-dot electrons. Both the *generation* and the *detection* of spin polarization on the quantum dot occur via electrical transport as a consequence of spin-polarized charge currents from and to the leads. One of the intriguing features of a quantum-dot spin-valve device is the possibility to further *manipulate* the dot spin. This can be done directly via an externally applied magnetic field [12]. But also the gate and transport voltages influence the dot spin [13–17]. To understand this, it is important to notice that the strong Coulomb interaction on the quantum dot yields many-body correlations. As for the spin degree of freedom, the quantum-dot electrons are subject to an exchange field that arises as a many-body effect due to the tunnel coupling to spin-polarized leads. This exchange field sensitively depends on the system parameters such as the gate and bias voltage. The latter, therefore, provide suitable handles to manipulate the quantum-dot spin.

Combining ferromagnetic (typically metallic) leads with quantum dots, which are usually semiconductor structures, is experimentally challenging. Recent experimental approaches to such a quantum-dot spin valve involve metallic islands [18, 19], granular systems [20, 21], carbon nanotubes [22–24] as well as single molecules [25] or self-assembled quantum dots [26, 27] coupled to ferromagnetic leads. Another possible realization would rely on contacting a surface impurity acting as quantum dot with a spin-polarized STM tip [28].

Successful demonstration of tunnel magnetoresistance through a strongly interacting system has been reported by Sahoo et al. [29] in single-walled carbon nanotubes contacted by PdNi leads, and by Zhang et al. in Al grains sandwiched inside a tunnel junction between two Co leads [30].

The article is organized as follows: In Sect. 6.2 we define the Hamiltonian of the quantum dot coupled to ferromagnetic leads. In Sect. 6.3 we address the dynamics of the dot spin and charge. Starting from a rigid calculation of the spin and charge current through a tunnel junction, we construct the master/Bloch equation for the charge/spin degrees of freedom from the charge/spin continuity equation. From the Bloch equation, we discuss, how to prepare, and modify the dot spin via bias voltage, gate voltage and an external applied magnetic field. The so prepared dot spin can be measured by its imprint on the conductance of the quantum dot spin-valve device as shown in Sect. 6.4. We summarize our findings in Sect. 6.5.

6.2 Model Hamiltonian

We describe the quantum-dot spin valve by the following Hamiltonian [13, 14]:

$$H = \sum_{rk\alpha} \varepsilon_{rk\alpha} c^{\dagger}_{rk\alpha} c_{rk\alpha} + \sum_{n} \varepsilon_n d^{\dagger}_n d_n + U d^{\dagger}_{\uparrow} d_{\uparrow} d^{\dagger}_{\downarrow} d_{\downarrow} + \sum_{rk\alpha n} \left(V_{rk\alpha n} c^{\dagger}_{rk\alpha} d_n + h.c. \right). \tag{6.1}$$

The first term in (6.1) treats the ferromagnetic leads $r = L/R$ as large reservoirs of itinerant electrons. The Fermion creation and annihilation operators of the lead r are labeled by $c^{(\dagger)}_{rk\alpha}$, where k labels the momentum and $\alpha = \pm$ the spin. The spin-quantization axis for the electrons in reservoir r is chosen along its magnetization direction $\mathbf{m}_r$. In the spirit of the Stoner model, the property of ferromagnetism is incorporated by assuming an asymmetry in the density of states ξ_α for majority (+) and minority (−) spins. The degree of spin polarization in lead r is characterized by the ratio $P_r = (\xi_{r+} - \xi_{r-})/(\xi_{r+} - \xi_{r-})$. The lead magnetization directions $\mathbf{m}_{\mathrm{L}}$ and $\mathbf{m}_{\mathrm{R}}$ can enclose an arbitrary angle ϕ. Furthermore, the leads shall be so large, that the electrons can always be described as in equilibrium, i.e., with a Fermi distribution $f_r(\omega)$. An applied bias voltage is taken into account by a symmetric shift of the chemical potential in the left and right lead by $\pm eV/2$.

The quantum dot can be modeled as an Anderson impurity with a single orbital level, where $d^{\dagger}_n$ and d_n are the Fermion creation and annihilation operators of the dot electrons with the spin $n = \uparrow, \downarrow$. The spin quantization axis of the dot is, in general, chosen to be different from the quantization axes of both the left and the right lead. If an external magnetic field is applied, the spin quantization axis is chosen parallel to this field. The energy ε_n of the atomic-like electronic level is measured relative to the equilibrium Fermi energy of the leads, and double occupation of the dot costs the charging energy $U \gg k_{\mathrm{B}}T$.

Electron tunneling between the leads and the dot is described by the last term in (6.1). As we have chosen different spin quantization axes for the lead subsystems, parallel to the respective magnetization, the tunneling matrix elements $V_{rk\alpha n}$ are not diagonal in spin space. However, we require that tunneling is spin conserving. The tunneling amplitudes can, then, be separated in $V_{rk\alpha n} = t_{rk} \times U^r_{\alpha n}$, i.e., a spin-independent tunnel amplitude t_{rk} and a $SU(2)$ rotation matrix $U^r_{\alpha n}$. The explicit shape of the matrix is determined by the choice of the dot spin quantization axis of the dot system.

The tunnel-coupling strength is characterized by the transition rates $\Gamma_{r\alpha}(\omega) = 2\pi \sum_k |t_{rk}|^2 \delta(\omega - \varepsilon_{rk\alpha})$. For simplicity, we assume the density of states ξ_α and the tunneling amplitudes t_r to be independent of energy, which implies constant tunneling rates $\Gamma_{r\alpha}$. The spin asymmetry in the density of states in the leads yields spin-dependent tunneling rates, which are related to the leads' spin polarization by $P_r = (\Gamma_{r+} - \Gamma_{r-})/(\Gamma_{r+} - \Gamma_{r-})$

Throughout this article we focus on the limit of dot-lead coupling $\Gamma_{r\alpha} \ll k_B T,\ eV$ when transport is dominated by first-order tunneling, i.e., it is sufficient to calculate all expressions for the charge and spin current up to first order in Γ. This excludes the regimes of second-order transport (cotunneling) in the Coulomb-blockade region [31, 32] or the Kondo regime (see, e.g., Ref. [33, 34]).

6.3 Quantum-Dot-Spin Dynamics

The tunnel coupling of the quantum-dot levels to spin-polarized leads yields a transfer of angular momentum across each of the tunnel junctions. This, together with a change of angular momentum due to an externally applied magnetic field, defines the dynamics of the quantum-dot spin polarization. The stationary value of the latter is determined by balancing all currents of angular momentum. As we discuss in detail below, the total *spin* current, i.e., transfer of angular momentum from the leads to the dot, consists of two qualitatively different contributions. One is associated with the fact that *charge* currents from or to a ferromagnet is spin polarized. This current, thus, transfers angular momentum that is oriented *along* either the magnetization direction of one of the ferromagnets or the direction of the accumulated quantum-dot spin. There is, however, also an additional contribution of transfer of *perpendicular* angular momentum, which can be expressed in terms of a many-body exchange field acting on the quantum-dot electrons.

For a careful treatment of the total transfer of angular momentum, we first present a rigid calculation of the spin current through a single tunnel junction [35] in terms of non-equilibrium Keldysh Green's functions. This will help us to identify under which circumstances spin currents with perpendicular components will contribute. Afterwards, we specify our result to the weak-coupling regime of a quantum-dot spin valve and derive in this limit Bloch-like rate equations for the quantum-dot spin.

The currents will be functions of the unknown density matrix elements. To derive the stationary density matrix in a non-equilibrium situation, we set the change of charge and spin on the dot equal zero. Then, the Bloch equation for the spin and the continuity equation for the charge degree of freedom from a system of master equations.

6.3.1 *Spin Current Expressed by Means of Green's Functions*

Our calculation of the spin current in this subsection will be in close analogy to the derivation of the charge current according to Meir and Wingreen [36]. Let us first consider the spin current through one, say the left, tunnel barrier. For a clearer notation, we mostly drop the lead index in this section. The spin current $\mathbf{J}_\mathrm{L} = \langle \hat{\mathbf{J}}_\mathrm{L} \rangle$ from the lead into the dot is defined by the negative of the time derivative of the total lead spin $\hat{\mathbf{S}}_\mathrm{L} = (\hbar/2) \sum_{k\alpha\beta} c^\dagger_{k\alpha} \boldsymbol{\sigma}_{\alpha\beta} c_{k\beta}$, where $\boldsymbol{\sigma}_{\alpha\beta}$ denotes the vector of Pauli matrices. From the Heisenberg equation we get

$$\hat{\mathbf{J}}_\mathrm{L} = -\frac{d}{dt}\hat{\mathbf{S}}_\mathrm{L} = -\frac{1}{\mathrm{i}\hbar}\Big[\hat{\mathbf{S}}_\mathrm{L}, H\Big]. \tag{6.2}$$

Making use of the Fermion commutation relation, we can find the spin-current operator as

$$\hat{\mathbf{J}}_\mathrm{L} = -\frac{1}{2\mathrm{i}} \sum_{k\alpha\beta n} V_{k\alpha n} \boldsymbol{\sigma}^\star_{\alpha\beta}\, c^\dagger_{k\beta}\, d_n - V^\star_{k\alpha n} \boldsymbol{\sigma}_{\alpha\beta}\, d^\dagger_n\, c_{k\beta}. \tag{6.3}$$

By introducing the Keldysh Green's functions $G_{n,k\beta}$ we can write the expectation value of the spin current as

$$\mathbf{J}_\mathrm{L} = \frac{1}{2}\sum_{k\alpha\beta n} \int \frac{d\omega}{2\pi}\Big(V_k \boldsymbol{\sigma}^\star_{\alpha\beta} G^<_{n,k\beta}(\omega) - V^\star_k \boldsymbol{\sigma}_{\alpha\beta} G^<_{k\beta,n}(\omega)\Big). \tag{6.4}$$

Since the Green's functions obey the Dyson equations $G^<_{k\alpha,n} = \sum_m V_{k\alpha,m}[g^t_{k\alpha} G^<_{m,n} - g^<_{k\alpha} G^{\bar{t}}_{m,n}]$ and $G^<_{n,k\alpha} = \sum_m V^\star_{k\alpha,m}[g^<_{k\alpha} G^t_{n,m} - g^{\bar{t}}_{k\alpha} G^<_{n,m}]$, we can replace the Green's functions in (6.4) with the dot Green's functions $G^<_{n,m}(t) = \mathrm{i}\langle d^\dagger_m d_n(t)\rangle$ and the free Green's functions of the lead. The latter are given by $g^<_{k\alpha} = 2\pi\mathrm{i} f^+_\mathrm{L}(\omega)\ \ \delta(\omega - \varepsilon_{k\alpha})$, $g^>_{k\alpha} = -2\pi\mathrm{i} f^-_\mathrm{L}(\omega)\delta(\omega - \varepsilon_{k\alpha})$, $g^\mathrm{ret}_{k\alpha} = 1/(\omega - \varepsilon_{k\alpha} + \mathrm{i}0^+)$, and $g^\mathrm{adv}_{k\alpha} = \left(g^\mathrm{ret}_{k\alpha}\right)^\star$. Here, f^+_L stands for the Fermi distribution function in the lead L and $f^-_\mathrm{L} = 1 - f^+_\mathrm{L}$.

If we choose the dot spin quantization axis parallel to the lead magnetization we can substitute the tunnel matrix elements by $V_{k\alpha,n} = t_k \delta_{\alpha n}$. After a lengthy but straightforward calculation, the spin current can be written as

$$\begin{aligned}\mathbf{J}_\mathrm{L} = & \frac{\mathrm{i}}{4}\sum_{m,n} \int \frac{d\omega}{2\pi}\, \boldsymbol{\sigma}_{mn}(\Gamma_m + \Gamma_n)\Big[f^+_\mathrm{L}(\omega) G^>_{n,m} + f^-_\mathrm{L}(\omega) G^<_{n,m}\Big] \\ & + \boldsymbol{\sigma}_{mn}(\Gamma_m - \Gamma_n)\left[f^+(\omega)(G^\mathrm{ret}_{n,m} + G^\mathrm{adv}_{n,m}) + \frac{1}{\mathrm{i}\pi}\int' dE \frac{G^<_{n,m}(E)}{E - \omega}\right],\end{aligned} \tag{6.5}$$

with the tunnel rates $\Gamma_n(\omega) = 2\pi \sum_k |t_{rk}|^2 \delta(\omega - \varepsilon_{rk\alpha})\delta_{\alpha n}$.

This is the most general expression for the spin current flowing through a tunnel barrier. Since the Green's functions $G_{n,m}$ were not specified during the calculation, (6.5) holds also for other electronic systems than single-level quantum dots.

If the dot state is rotationally symmetric about $\mathbf{m}_\mathrm{L}$, all dot Green's functions $G_{\sigma\sigma'}$ non-diagonal in spin space vanish. Only in this case, the spin current is proportional to the difference between charge current $I_{\mathrm{L}}^{\uparrow} = \mathrm{i}(e/h) \int d\omega \Gamma_{\uparrow} [f_{\mathrm{L}}^{+}(\omega) G_{\uparrow\uparrow}^{>} + f_{\mathrm{L}}^{-}(\omega) G_{\uparrow\uparrow}^{<}]$ carried by spin-up electrons and charge current $I_{\mathrm{L}}^{\downarrow}$ carried by spin-down electrons,

$$\mathbf{J}_{\mathrm{L}} = J_{\mathrm{L}}^{z} \mathbf{e}_{z} = \frac{\hbar}{2e} \left(I_{\mathrm{L}}^{\uparrow} - I_{\mathrm{L}}^{\downarrow} \right) \mathbf{e}_{z}. \tag{6.6}$$

If the dot system breaks this rotational symmetry, for example due to spin accumulation along an axis different from $\mathbf{m}_{\mathrm{L}}$, the simple result of (6.6) is no longer correct. In such a situation, the second line in (6.5) yields an additional spin-current component, oriented transversal to both, the magnetization of the lead, and the polarization of the dot. This spin-current component describes the exchange coupling between lead and dot spin, causing both to precess around each other. Since the lead magnetization is pinned usually, only the dot spin precesses like in a magnetic field.

Brataas et al. [37] showed, that at normal metal–ferromagnet interfaces, incoming electrons, with a spin orientation non-collinear to the magnetization direction, may experience a rotation of the spin direction during backscattering. This mechanism is described by the so called spin-mixing conductance, and also generates a transverse component of the spin current.

6.3.2 *Spin Current Between Ferromagnetic Lead and Quantum Dot*

We now specify the above expressions for a quantum-dot spin valve for weak tunnel coupling. Since the expression for the spin current in (6.5) does already explicitly depend linearly on the tunnel coupling $\Gamma_{r\sigma}$, we only need the zeroth-order Keldysh Green's functions of the dot system to describe the weak-coupling regime. They are given by

$$G_{\sigma\sigma}^{>}(\omega) = -2\pi \mathrm{i} P_{\bar{\sigma}} \delta(\omega - \varepsilon - U) - 2\pi \mathrm{i} P_{0} \delta(\omega - \varepsilon) \tag{6.7}$$

$$G_{\sigma\sigma}^{<}(\omega) = 2\pi \mathrm{i} P_{\sigma} \delta(\omega - \varepsilon) + 2\pi \mathrm{i} P_{d} \delta(\omega - \varepsilon - U) \tag{6.8}$$

$$G_{\sigma\bar{\sigma}}^{>}(\omega) = 2\pi \mathrm{i} P_{\bar{\sigma}}^{\sigma} \delta(\omega - \varepsilon - U) \tag{6.9}$$

$$G_{\sigma\bar{\sigma}}^{<}(\omega) = 2\pi \mathrm{i} P_{\bar{\sigma}}^{\sigma} \delta(\omega - \varepsilon) \tag{6.10}$$

$$G_{\sigma\bar{\sigma}}^{\mathrm{ret}}(\omega) = \frac{P_{\bar{\sigma}}^{\sigma}}{\omega - \varepsilon + \mathrm{i}0^{+}} + \frac{P_{\bar{\sigma}}^{\sigma}}{\omega - \varepsilon + U + \mathrm{i}0^{+}} = \left(G_{\sigma\bar{\sigma}}^{\mathrm{adv}}(\omega) \right)^{\star} \tag{6.11}$$

where P_{η}^{χ} are the matrix elements of the reduced density matrix of the dot system,

$$\rho_{\text{dot}} = \begin{pmatrix} P_0^0 & 0 & 0 & 0 \\ 0 & P_\uparrow^\uparrow & P_\downarrow^\uparrow & 0 \\ 0 & P_\uparrow^\downarrow & P_\downarrow^\downarrow & 0 \\ 0 & 0 & 0 & P_{\text{d}}^{\text{d}} \end{pmatrix}. \tag{6.12}$$

The diagonal, real entries $P_\chi^\chi \equiv P_\chi$ are the probabilities to find the dot in the state empty (0), occupied with a spin up $(\uparrow)$ or down $(\downarrow)$ electron, or doubly occupied (d) with a spin singlet. The zeros in (6.12) in the off diagonals are a consequence of the total-particle-number conservation. The inner 2×2 matrix is the $SU(2)$ representation of the dot spin. The reduced density matrix contains five independent parameters. For convenience, we describe the quantum dot state by the probabilities for the three charge states P_0, $P_1 = P_\uparrow^\uparrow + P_\downarrow^\downarrow$, and P_{d} (with the normalization condition $P_0 + P_1 + P_{\text{d}} = 1$), and the average-spin vector $\mathbf{S} = (P_\downarrow^\uparrow + P_\uparrow^\downarrow, \mathrm{i}P_\downarrow^\uparrow - \mathrm{i}P_\uparrow^\downarrow, P_\uparrow^\uparrow - P_\downarrow^\downarrow)/2$.

Similarly to deriving the expression for the spin current, we can get a formula for the charge current I_r through tunnel contact r as

$$I_r = -\frac{e}{h}\sum_{k\alpha n}\int d\omega\left(V_{rk\alpha n}G^<_{n,rk\alpha}(\omega) - V^\star_{rk\alpha n}G^<_{rk\alpha,n}(\omega)\right). \tag{6.13}$$

After choosing a spin quantization axis for the dot spin and making use of the Dyson equation for the Green's functions, we can plug in the dot Green's function given above to obtain the result

$$\begin{aligned} I_r = \Gamma_r \frac{2(-e)}{\hbar}\Bigg[& f_r^+(\varepsilon)P_0 + \frac{f_r^+(\varepsilon + U) - f_r^-(\varepsilon)}{2}P_1 - f_r^-(\varepsilon + U)P_{\text{d}} \\ & - p_r\left[f_r^-(\varepsilon) + f_r^+(\varepsilon + U)\right]\mathbf{S}\cdot\mathbf{m}_r\Bigg], \end{aligned} \tag{6.14}$$

that, of course, is independent of the choice of the dot spin's quantization axis. Here, we defined $\Gamma_r \equiv (\Gamma_{r\uparrow} + \Gamma_{r\downarrow})/2$.

It is worth to mention that the dot spin $\mathbf{S}$ influences the conductance via the scalar product $(\mathbf{S}\cdot\mathbf{m}_r)$. Therefore the tunnel magnetoresistance depends cosine-like on the relative angle enclosed by lead magnetization and spin polarization, i.e., it just resembles the behavior of a tunnel junction between two ferromagnetic leads [8–11].

The first-order spin current, on the other hand, is given by evaluating (6.5), which leads to

$$\mathbf{J}_r = \frac{\hbar}{2e}I_r p_r \mathbf{m}_r - \frac{\mathbf{S} - p_r^2(\mathbf{m}_r\cdot\mathbf{S})\mathbf{m}_r}{\tau_{\text{c},r}} + \mathbf{S}\times\mathbf{B}_r. \tag{6.15}$$

The first term describes spin injection from the ferromagnetic lead into the quantum dot by a spin-polarized charge current. The injected spin is proportional to the lead polarization and the electrical current crossing the junction. This spin current contribution vanishes for vanishing bias voltage.

The second term describes relaxation of the dot spin due to coupling to the leads. Since neither an empty nor a doubly occupied dot can bear a net spin, the spin relaxation time $\tau_{c,r}^{-1} = \Gamma_r/\hbar(1 - f_r(\varepsilon) + f_r(\varepsilon + U))$ equals the life time of the single-occupation dot state. This relaxation term is anisotropic [38]. The spin polarization of the lead suppresses the relaxation of a dot spin, which is aligned collinear to the lead magnetization.

The third term in (6.15) describes transfer of angular momentum perpendicular to the spin-polarization directions of lead and dot. The structure of this terms suggests the interpretation of $\mathbf{B}_r$ as being an effective magnetic field that acts on the quantum-dot spin $\mathbf{S}$. Its value, in the absence of an external magnetic field, is given by [13, 39]

$$\mathbf{B}_r = p_r \frac{\Gamma_r \mathbf{m}_r}{\pi\hbar} \int' d\omega \left(\frac{f_r^+(\omega)}{\omega - \varepsilon - U} + \frac{f_r^-(\omega)}{\omega - \varepsilon} \right), \tag{6.16}$$

where the prime at the integral indicates Cauchy's principal value. From (6.16) it is clear that this field is an exchange field that arises due to the fact that the quantum dot levels are tunnel-coupled to a spin-polarized lead. It is a many-body effect as all degrees of freedom in the leads contribute to the integral, and the Coulomb interaction in the dot is important not to cancel the first with the second term in the integrand. The exchange field persists also for vanishing bias voltage. A signature of this exchange field in the Kondo-resonance splitting of transport through a single molecule has been observed recently [25], with reported values of the field of up to 70 T.

6.3.3 Dynamics of the Quantum-Dot Spin

We use the expressions for the charge and spin current, (6.14) and (6.15), to calculate the dynamics of the dot's charge and spin. Strictly speaking, the above calculation holds only for static systems. To emphasize the physical origin of the following equations, we keep all time derivatives in this section, even if they should have the numerical value of zero.

The continuity equation of the average dot charge $\langle n \rangle = \sum_n nP_n$ is given by

$$e\frac{d\langle n \rangle}{dt} = I_\mathrm{L} + I_\mathrm{R}, \tag{6.17}$$

see Fig. 6.2a. Moreover, not only the total charge current through the dot is conserved, but also the charge current through the individual charge levels. Therefore we can split the charge continuity (6.17) into the two contributions

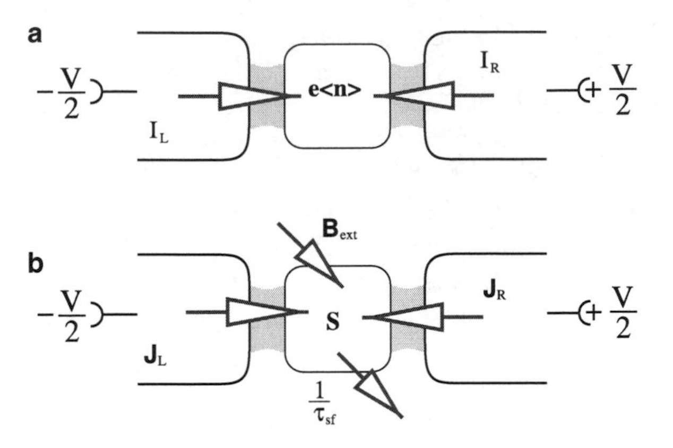

Fig. 6.2 **a** The dot charge changes according to the electrical current through the tunnel barriers. **b** The dot spin changes according to the spin currents through the tunnel barriers. In addition, an external magnetic field acts as additional source and the intrinsic spin relaxation as sink of angular momentum

associated with transport processes in which either double occupied or an empty dot is involved. The affiliation to either contribution is indicated by the arguments of the Fermi functions, where the presence of the interaction energy U indicates processes with double occupation and the absence signals processes involving an empty dot. We get

$$\begin{aligned}\frac{dP_0}{dt} &= \sum_r \Gamma_r \big(f_r^+(\varepsilon)P_0 - f_r^-(\varepsilon)P_1/2 - p_r f_r^-(\varepsilon)\mathbf{S}\cdot\mathbf{m}_r\big)\\ \frac{dP_{\mathrm{d}}}{dt} &= \sum_r \Gamma_r \big(f_r^+(\varepsilon+U)P_1/2 - f_r^-(\varepsilon+U)P_{\mathrm{d}} - p_r f_r^+(\varepsilon+U)\mathbf{S}\cdot\mathbf{m}_r\big).\end{aligned} \tag{6.18}$$

Similar to the charge continuity equation, the continuity equation for the dot spin, see Fig. 6.2b, reads

$$\begin{aligned}\frac{d\mathbf{S}}{dt} &= \mathbf{J}_{\mathrm{L}} + \mathbf{J}_{\mathrm{R}} + \mathbf{S}\times\mathbf{B}_{\mathrm{ext}} - \frac{\mathbf{S}}{\tau_{\mathrm{rel}}}\\ &= \frac{\hbar}{2e}\sum_r \left[I_r p_r \mathbf{m}_r - \frac{\mathbf{S} - p^2(\mathbf{m}_r\cdot\mathbf{S})\mathbf{m}_r}{\tau_{\mathrm{c},r}}\right] + \mathbf{S}\times\mathbf{B}_{\mathrm{tot}} - \frac{\mathbf{S}}{\tau_{\mathrm{rel}}}.\end{aligned} \tag{6.19}$$

with $\mathbf{B}_{\mathrm{tot}} = (\mathbf{B}_{\mathrm{L}} + \mathbf{B}_{\mathrm{R}} + \mathbf{B}_{\mathrm{ext}})$. In addition to the spin currents entering the quantum dot from the left and right lead, there is one term describing spin precession due to an external magnetic field $\mathbf{B}_{\mathrm{ext}}$. It enters the equation in the same way as the exchange field with the left and right reservoir, so that all three of them add up to the total field $\mathbf{B}_{\mathrm{tot}}$. Furthermore, we phenomenologically took into account the possibility of intrinsic spin relaxation, e.g., due to spin-orbit coupling, hyperfine interaction with nuclei in the quantum dot, or higher-order tunnel processes such as spin-flip cotunneling, with a time scale τ_{rel}. The total spin-decoherence time of the dot spin is, therefore, given by

$$(\tau_{\mathrm{s}})^{-1} = (\tau_{\mathrm{rel}})^{-1} + (\tau_{\mathrm{c,L}})^{-1} + (\tau_{\mathrm{c,R}})^{-1}. \tag{6.20}$$

The different handles to manipulate the quantum-dot spin are comprised in the total field $\mathbf{B}_{\mathrm{tot}}$. It contains the external magnetic field $\mathbf{B}_{\mathrm{ext}}$ as a direct tool to initiate a spin precession. However, also the exchange fields can be used for this task [14]. As we see from (6.16), the exchange field depends on both the gate and bias voltage via the level position ε and the Fermi distribution functions $f_r(\omega)$.

6.4 Manipulation and Detection of the Quantum-Dot Spin via Electrical Transport

Since the spin state of the quantum dot enters the expressions for the charge current in (6.14), any manipulation on the quantum-dot spin can be detected in measuring the *dc*-charge current through the device.

In order to calculate the charge current, we need to determine the stationary solution for the density matrix, i.e., for the dot spin $\mathbf{S}$ and the charge occupation probabilities P_i. For these six variables, we need six independent equations: the probability normalization condition $\sum_n P_n = 1$, the Bloch equation $dS/dt = 0$ (contains three equations), and the two equations originating from the charge continuity.

In the following three subsections we consider the effect of the gate and transport voltage as well as an external magnetic field on the quantum-dot spin, and how this is reflected by electric transport through the quantum-dot spin valve. Then, in reversal, by experimentally measuring the transport characteristics of the device, one can conclude the spin state of the quantum dot.

In the stationary transport situation under consideration, neither the average charge nor the average spin of the dot change with time, and the currents through the left and right tunnel junction are equal $I_\mathrm{L} = -I_\mathrm{R} \equiv I$. In order to simplify the following discussion, we choose symmetric coupling $\Gamma_L = \Gamma_R = \Gamma/2$, equal spin polarizations $p_\mathrm{L} = p_\mathrm{R} = p$, and a symmetrically applied bias $V_R = -V_L = V/2$.

6.4.1 Gate Voltage Effect in the Linear-Response Regime

To study the effect of the gate voltage on the quantum-dot spin via the gate-voltage dependence of the exchange field, we analyze the linear-response regime in the absence of an external magnetic field. Without any applied bias voltage $V = 0$, i.e., in equilibrium, the stationary solution of the rate equations for the charge occupation probabilities (6.18) is given by the Boltzmann distribution, $P_x \sim exp(-E_\chi/k_\mathrm{B}T)$, and no current flows. Since the dot itself is non-magnetic, the dot spin vanishes $\mathbf{S} = 0$. For a small bias voltage $eV \ll k_\mathrm{B}T$, we can expand the master equation (6.19) and (6.18) up to linear order in V. With symmetric couplings to the left and right lead, the charge probabilities (P_0, P_1, P_d) become independent of V, thus, the occupation probabilities are given by their equilibrium value, but the spin degree of freedom is not. In linear response, the linear charge current, $I_\mathrm{lin} = V \cdot \partial I/\partial V|_{\mathrm{V}=0}$, which is polarized due to the lead magnetizations generate a

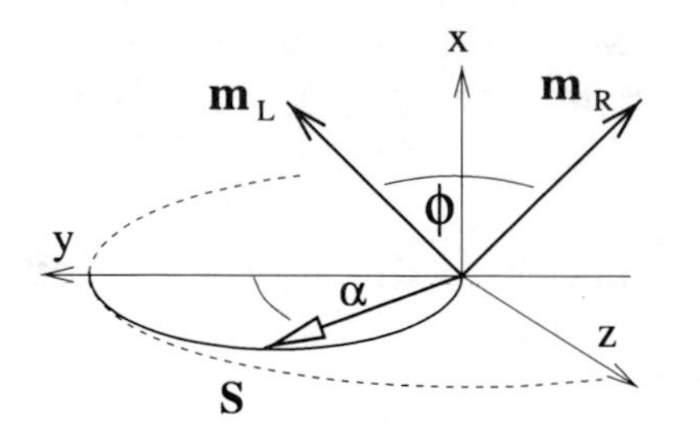

Fig. 6.3 Spin dynamics in the linear-response regime. Spin accumulates along the y direction. The spin precesses due to the exchange field that is along the x direction. Therefore, the stationary solution of the average spin on the dot is tilted away from the y axis by an angle α, plotted in Fig. 6.4b

finite dot-spin polarization along $p(\mathbf{m}_L - \mathbf{m}_R)$, i.e., along the y axis in the coordinate system defined in Fig. 6.3.

The damping term in (6.19) limits the magnitude of spin accumulation. The term $S \times (B_L + B_R)$ yields an intrinsic precession of the spin around the exchange field $B_L + B_R \equiv B_0 \cos(\phi/2)\mathbf{e}_x$. In the steady state, the average dot spin is rotated by the angle

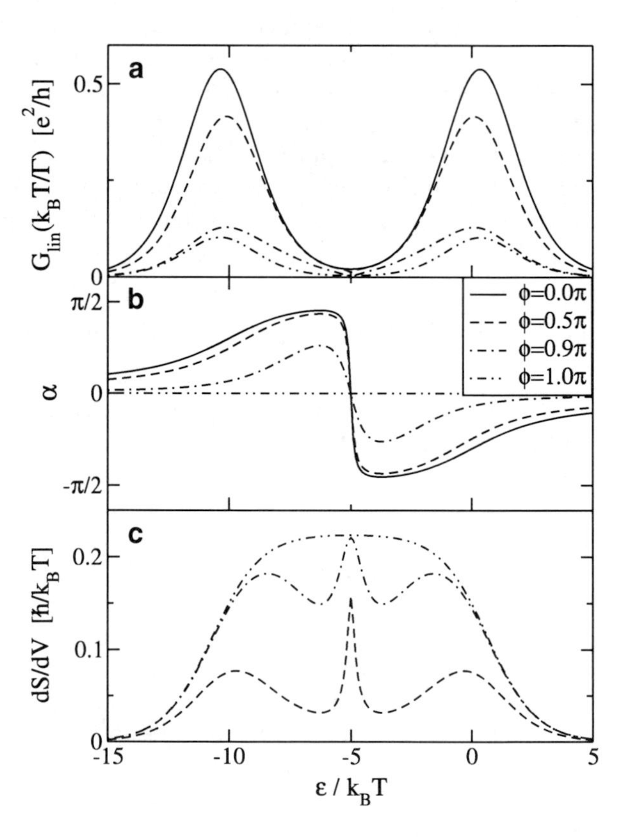

Fig. 6.4 **a** Linear conductance normalized by Γ/k_BT as a function of the level position ε for different angles ϕ. **b** Angle α enclosed by the accumulated spin and the y axis as defined in Fig. 6.3. **c** Derivation of the magnitude of the accumulated spin on the dot with respect to the source-drain voltage V. Further parameters are $p = 0.9$, $\tau_{rel}^{-1} = 0$, and $U = 10\, k_BT$

$$\alpha = -\arctan\left(B_0\tau_s\cos\frac{\phi}{2}\right). \tag{6.21}$$

Therefore the accumulated spin acquires both y and z components as seen in Fig. 6.3. This precession also leads to a reduction of the magnitude of the accumulated spin.

The precession angle α is plotted in Fig. 6.4b as function of the level position ε, that can be tuned by the gate voltage. The angle α changes its sign at $\varepsilon = -U/2$, due to a sign change of the exchange field at this point. The level position $\varepsilon = -U/2$ is special, since then the particle- and hole-like processes generating the exchange field compensate each other.

As pointed out above, in the linear-response regime under consideration the charge occupation probabilities do not depend on the spin polarization of the leads. In particular, they are independent of the relative angle ϕ of the leads' magnetization. This means that the ϕ dependence of the conductance is determined by the product $\mathbf{S}\cdot\mathbf{m}_L = -\mathbf{S}\cdot\mathbf{m}_R$, as can be seen from (6.14). It is the relative orientation of the accumulated spin and the drain (or source) that produces the ϕ dependence of the current, rather than the product $\mathbf{m}_L\cdot\mathbf{m}_R$, as in the case of a single magnetic tunnel junction. Therefore, the ϕ dependent linear conductance $G_{\text{lin}} = (\partial I/\partial V)|_{V=0}$ directly reflects the accumulated spin. The effect of the exchange field for the normalized conductance is seen from the analytic expression

$$\frac{G_{\text{lin}}(\phi)}{G_{\text{lin}}(0)} = 1 - p^2\frac{\tau_s}{\tau_c}\frac{\sin^2(\phi/2)}{1+(B_0\tau_s)^2\cos^2(\phi/2)}, \tag{6.22}$$

which is plotted in Fig. 6.5 for different values of the level position ε.

For $\varepsilon > 0$, the quantum dot is predominantly empty, and for $\varepsilon + U < 0$ doubly occupied with a spin singlet. In these regions, the life time of a singly-occupied dot τ_c is short, and so is the lifetime of the dot spin. Therefore, the rotation angle α is small and the normalized conductance as a function of the relative angle ϕ of the lead magnetizations shows a harmonic behavior, see, e.g., the curve for $\varepsilon = 5$ k_BT in Fig. 6.5.

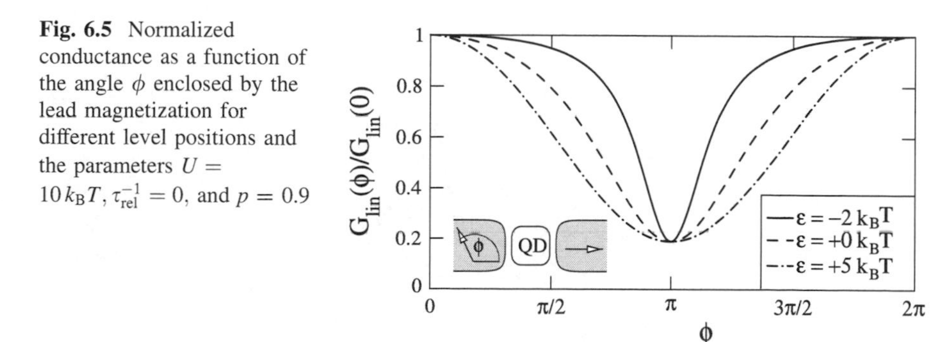

Fig. 6.5 Normalized conductance as a function of the angle ϕ enclosed by the lead magnetization for different level positions and the parameters $U = 10\,k_BT$, $\tau_{\text{rel}}^{-1} = 0$, and $p = 0.9$

For $-U < \varepsilon < 0$ the dot is primarily singly occupied, so the spin dwell time is increased and the exchange field becomes important. It causes the above described spin precession, which decreases the product $\mathbf{S} \cdot \mathbf{m}_{\mathrm{L}}$ since the relative angle between $\mathbf{m}_{\mathrm{L}}$ and $\mathbf{S}$ is increased and the magnitude of $\mathbf{S}$ is reduced. Thus, the spin precession makes the spin-valve effect less pronounced, leading to a value of the conductance that exceeds the expectations made by Slonczewski in Ref. [9] for a single magnetic tunnel junction.

For parallel and antiparallel aligned lead magnetizations, $\phi = 0$ and $\phi = \pi$, the accumulated spin and the exchange field also get aligned. In this case, the spin precession stops, even though the exchange field is still present. The ϕ-dependent conductance is not affected by the exchange field at this alignment, see Fig. 6.5.

6.4.2 Bias Voltage Effect in Non-Linear Regime

We now turn to the non-linear response regime, $eV > k_{\mathrm{B}}T$, in order to discuss the effect of the bias-voltage dependence of the exchange field on the quantum-dot spin. Again, we assume that there is no external magnetic field, and no spin relaxation.

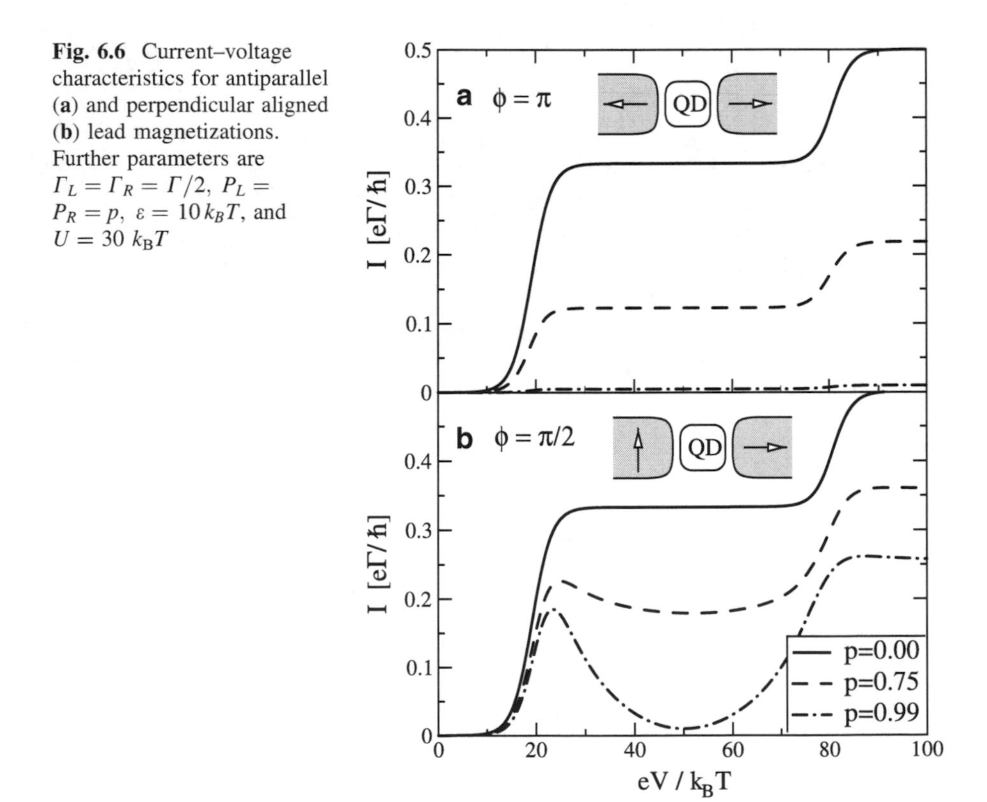

Fig. 6.6 Current–voltage characteristics for antiparallel (**a**) and perpendicular aligned (**b**) lead magnetizations. Further parameters are $\Gamma_L = \Gamma_R = \Gamma/2$, $P_L = P_R = p$, $\varepsilon = 10\,k_B T$, and $U = 30\ k_{\mathrm{B}}T$

In Fig. 6.6a we show the current I as a function of the bias voltage V for an antiparallel configuration of the leads' magnetizations and different values of the leads' spin polarization p.

For non-magnetic leads, the current–voltage characteristic shows the usual Coulomb staircase. At low bias voltage, the dot is empty and transport is blocked. With increasing bias voltage, first single and then double occupancy of the dot is possible, which opens first one and then two transport channels. A finite spin polarization p leads to spin accumulation and, thus, to a reduction of transport. A reduction of transport with increasing p is also seen for noncollinear magnetization. But there is a qualitative difference as can be seen in Fig. 6.6b. A very pronounced negative differential conductance evolves out of the middle plateau as p is increased. To understand the negative differential conductance we first neglect the exchange field and then, in a second step, analyze how the exchange field modifies the picture.

At the intermediate bias voltages ($\varepsilon \ll eV/2 \ll e + U$) the dot can only be empty or singly occupied. Since double occupation is forbidden, the transport through the dot is limited by the dot state with maximal dwell time in the dot. Due to the finite bias voltage, the dot electrons can only leave the dot to the drain (right) lead. Therefore, the electron with the longest dwell time is the one with its spin polarized antiparallel to the drain lead magnetization direction. For this antiparallel spin alignment the tunneling to the drain lead is maximally suppressed, while the tunneling from the source lead is not as much affected. When the tunneling to the drain lead is weak, but strong to the source lead, then the dot is primarily occupied by one electron (charge accumulation). The spin accumulation as function of the charge occupation P_1 given by

$$\mathbf{S} = p\left[\frac{\Gamma_{\mathrm{L}}}{\Gamma_{\mathrm{R}}}(1 - P_1)\mathbf{m}_{\mathrm{L}} - \frac{P_1}{2}\mathbf{m}_{\mathrm{R}}\right], \tag{6.23}$$

directly relates an increased probability P_1 to find the dot occupied by one electron to an average dot spin $\mathbf{S}$, aligned nearly antiparallel to the drain lead.

The behavior is different from that in the linear-response regime, where the direction of the accumulated spin is along $\mathbf{m}_{\mathrm{L}} - \mathbf{m}_{\mathrm{R}}$ rather than $-\mathbf{m}_{\mathrm{R}}$. This is related to the fact that in the linear-response regime, the dot electrons can both tunnel to the source (left) and drain (right) lead (where tunneling to the drain is only somewhat more likely than tunneling to the source). The direction of the accumulated spin does, therefore, not only depend on the magnetization of the drain but also on that of the source electrode. To minimize transport, the accumulated spin will be aligned along $\mathbf{m}_{\mathrm{L}} - \mathbf{m}_{\mathrm{R}}$.

Due to the fact that double occupation is forbidden, $P_{\mathrm{d}} \approx 0$, all electrons entering the dot through the left barrier must find an empty dot, i.e., the current $I = (e\Gamma/\hbar)P_0$ explicitly depends only on the probability to find the dot empty. Therefore, any additional charge accumulation, $\sim P_1$, on the dot reduces the conductance of the device proportional to $P_0 = 1 - P_1$. So this mechanism is a type of spin blockade but with a different physical origin compared to the

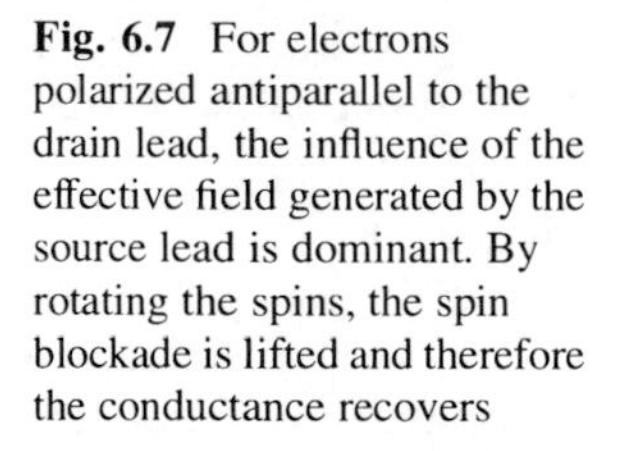

Fig. 6.7 For electrons polarized antiparallel to the drain lead, the influence of the effective field generated by the source lead is dominant. By rotating the spins, the spin blockade is lifted and therefore the conductance recovers

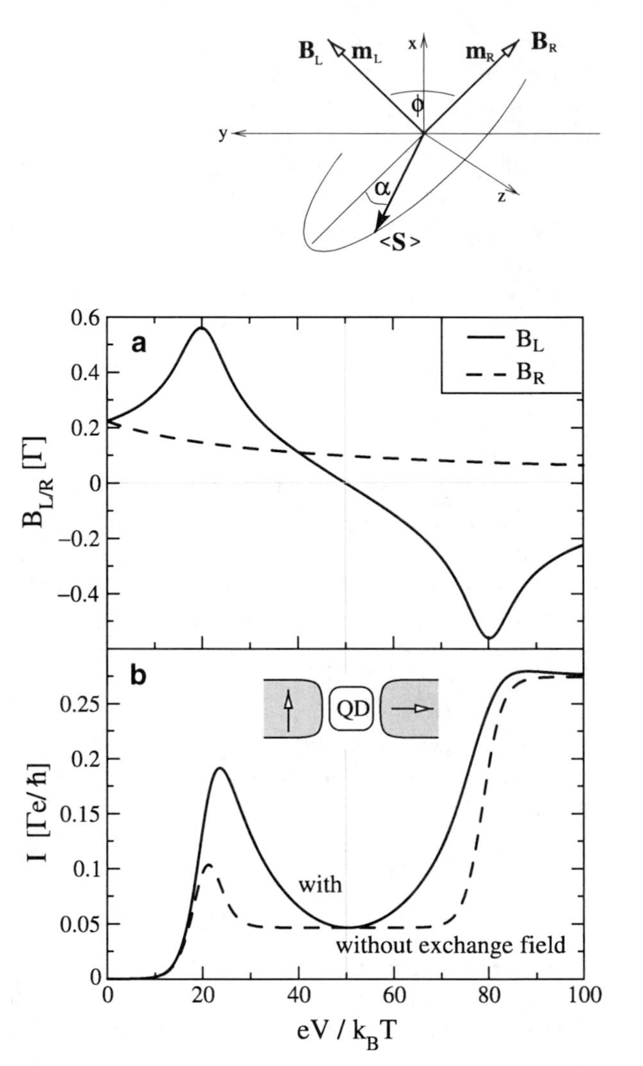

Fig. 6.8 **a** The absolute value of the effective exchange field contributions from the left and right leads. **b** The current–voltage dependence, with and without the influence of the exchange field. For both plots the parameters $\phi = \pi/2$, $\Gamma_L = \Gamma_R = \Gamma/2$, $\varepsilon = 10\ k_B T$, $U = 30\ k_B T$, and $p = 0.95$ were chosen

systems described in literature [40–42]. The suppression defines the local minimum of the current in Fig. 6.6b. At this point of the minimal current, the relevant exchange-field component generated by the coupling to the source (left) lead vanishes, see Fig. 6.8. Away from this point, the exchange field component perpendicular to the spin, (originating from the source lead) will induce a precession of this spin about $\mathbf{m}_L$ as illustrated in Fig. 6.7, and effectively diminish the spin blockade.

The particular value of the non-linear conductance is a consequence of the two competing effects. Spin blockade reduces, while spin precession, which reduces the spin blockade, again increases the conductance. Since the strength of the exchange field, which generates precession, varies as a function of bias voltage, see Fig. 6.8a, this recovery is non-monotonous function, what leads to a negative

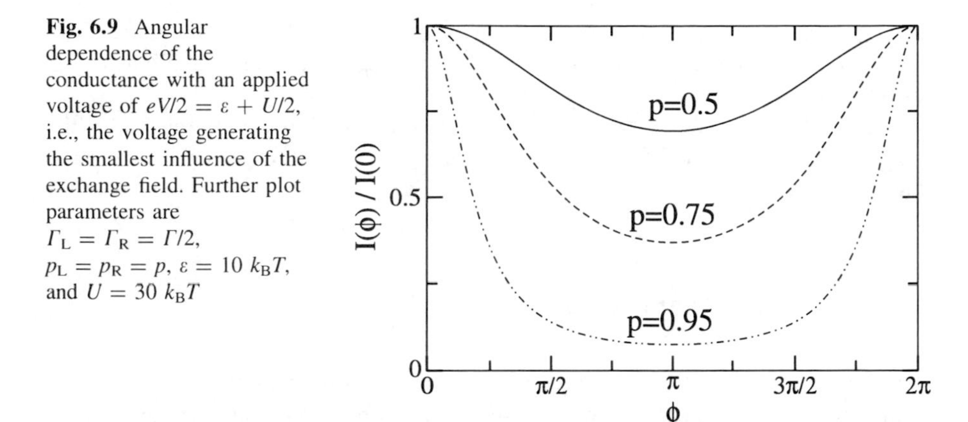

Fig. 6.9 Angular dependence of the conductance with an applied voltage of $eV/2 = \varepsilon + U/2$, i.e., the voltage generating the smallest influence of the exchange field. Further plot parameters are $\Gamma_\mathrm{L} = \Gamma_\mathrm{R} = \Gamma/2$, $p_\mathrm{L} = p_\mathrm{R} = p$, $\varepsilon = 10\ k_\mathrm{B}T$, and $U = 30\ k_\mathrm{B}T$

differential conductance. To illustrate this further, we plot in Fig. 6.8b the current which we obtained when the spin precession contribution has artificially been dropped in (6.19), and compare it with the total current. In the absence of the exchange field, a wide plateau is recovered, whose height is similar to the current one would expect if the lead magnetizations were aligned antiparallel. The peak at the left end of the plateau indicates that, once the dot level is close to the Fermi level of the source electrode, the spin blockade is relaxed since the dot electrons have the possibility to leave to the left side.

However, this negative differential conductance occurs only at relatively high values of the lead polarization. For symmetric tunnel coupling a spin polarization of $p \approx 0.77$ is needed, while for a strong asymmetry in the tunnel coupling the required spin polarization is reduced.

The effect of the spin blockade on the ϕ-dependence of the current is depicted in Fig. 6.9. We choose the bias voltage according to $eV/2 = \varepsilon + U/2$, such that the influence of the exchange field is absent. For $p = 0.5$ still a $\sin^2 \frac{\phi}{2}$ dependence can be recognized. For higher values of the spin polarization the conductance drops faster and stays nearly constant at its minimal value due to spin blockade. This is just the opposite behavior than predicted for the linear-response regime as seen in Fig. 6.5.

If such a high bias voltage is applied, that the dot can also be doubly occupied, the step-like behavior of the current–voltage characteristic is recovered, see Fig. 6.6b. Away from the step, all appearing Fermi functions can be approximated by 0 or 1, and following (6.14) the current is given by $I = (e\Gamma/2\hbar)$ $[1 - p\mathbf{S} \cdot (\mathbf{m}_\mathrm{L} - \mathbf{m}_\mathrm{R})]$. Far away from the resonance, where the exchange field can be neglected, the accumulated spin is $\mathbf{S} = p(\mathbf{m}_\mathrm{L} - \mathbf{m}_\mathrm{R})/4$ from which we get

$$I = \frac{e\Gamma}{2\hbar}\left(1 - p^2 \sin^2 \frac{\phi}{2}\right). \tag{6.24}$$

The suppression of transport due to the spin polarization p of the leads is comparable with the case of a single tunnel junction, when charging effects are of no importance.

We close this section with the remark that while we plotted only results for the case $\varepsilon > 0$, in the opposite case $\varepsilon < 0$ the current–voltage characteristics is qualitatively the same.

6.4.3 External Magnetic Field

In the previous subsections we studied quantum-dot spin dynamics evoked by the exchange field. But one can also make use of an externally-applied magnetic field $\mathbf{B}_{ext}$. It turns out that with an external field one can measure the spin-decoherence time T_2. To emphasize this point, we explicitly allow intrinsic spin relaxation on the dot. Then, we observe a separation of the inverse charge life time $\tau_c^{-1} = \tau_{c,L}^{-1} + \tau_{c,R}^{-1}$ and the inverse spin life time on the dot $\tau_s^{-1} = \tau_c^{-1} + \tau_{rel}^{-1}$.

An external field leads to the Hanle effect [12], i.e., the decrease of spin accumulation in the quantum dot due to precession about a static magnetic field. Indeed, this was the effect used by Johnson and Silsbee [43, 44] and others [19] to prove non-equilibrium spin accumulation.

Optical realizations of such Hanle experiments [45] always involve ensemble averaging over different dot realizations, so the outcome of the measurement is $T_2^\star$ rather than T_2. By measuring the Hanle signal via the conductance through a quantum dot attached to ferromagnetic leads, this ensemble averaging is avoided.

In a recent experiment Zhang et al. [30] realized this kind of setup but with a whole layer of aluminum dots in a tunnel junction between two Co electrodes. Although the measurements involve averaging over different realizations of the dots, multi levels and local magnetizations, they clearly observe a Hanle resonance in the magnetoresistance of the device.

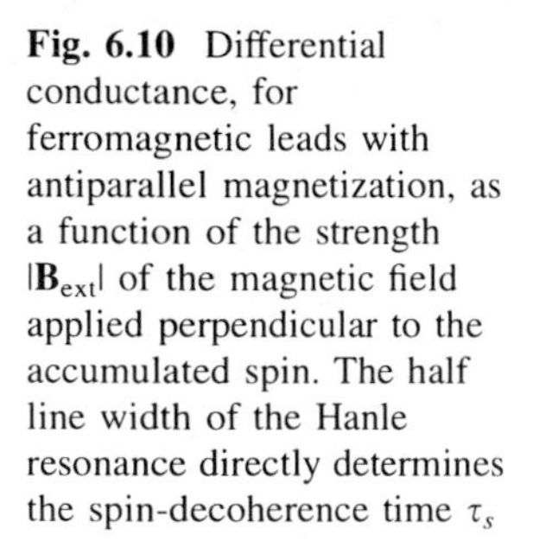

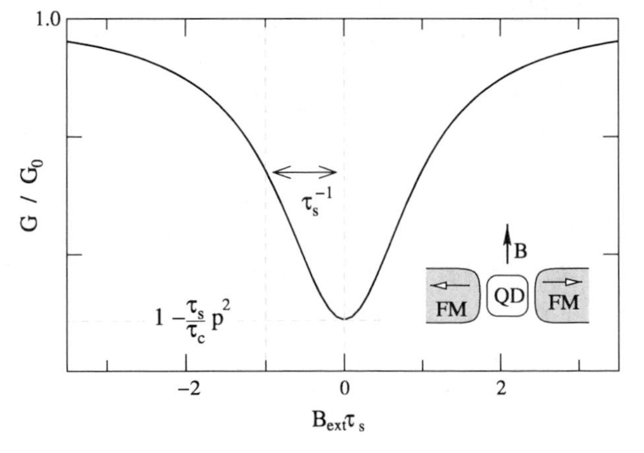

Fig. 6.10 Differential conductance, for ferromagnetic leads with antiparallel magnetization, as a function of the strength $|\mathbf{B}_{ext}|$ of the magnetic field applied perpendicular to the accumulated spin. The half line width of the Hanle resonance directly determines the spin-decoherence time τ_s

For simplicity we assume symmetric couplings $\Gamma_L = \Gamma_R$, equal degree of lead polarizations $p_L = p_R = p$ and consider the linear-response regime only. There is a variety of possible relative orientations of the external field and the leads' magnetizations to each other. In the following, we consider two specific cases in detail, as they are convenient to extract useful information about the spin-decoherence time in one case, and to prove the existence of the exchange field in the other.

6.4.3.1 Antiparallel Aligned Leads' Magnetizations

We first focus on two ferromagnetic leads with magnetization directions antiparallel to each other, see Fig. 6.10, and an arbitrary aligned external field. The configuration has the advantage that the exchange field contributions from the two leads cancel, and the spin dynamics is only governed by the external field $\mathbf{B}_{\text{ext}}$.

The linear conductance, then, is

$$\frac{G}{G_0} = 1 - p^2 \frac{\tau_s}{\tau_c} \frac{1 + \left(\frac{\mathbf{m}_L - \mathbf{m}_R}{2} \mathbf{B}_{\text{ext}} \tau_s\right)^2}{1 + (\mathbf{B}_{\text{ext}} \tau_s)^2}. \tag{6.25}$$

where $G_0 = e^2 P_1/\tau_c k_B T$ is the asymptotic value of the conductance for a large magnetic field, $|\mathbf{B}_{\text{ext}}| \to \infty$, for which the spin accumulation is completely destroyed. The latter is proportional to the single occupation probability P_1.

If we assume the field to be aligned perpendicular to the lead magnetizations (see Fig. 6.10), we find the Lorentzian dependence on the external magnetic field that is familiar from the optical Hanle effect. The depth of the dip is given by

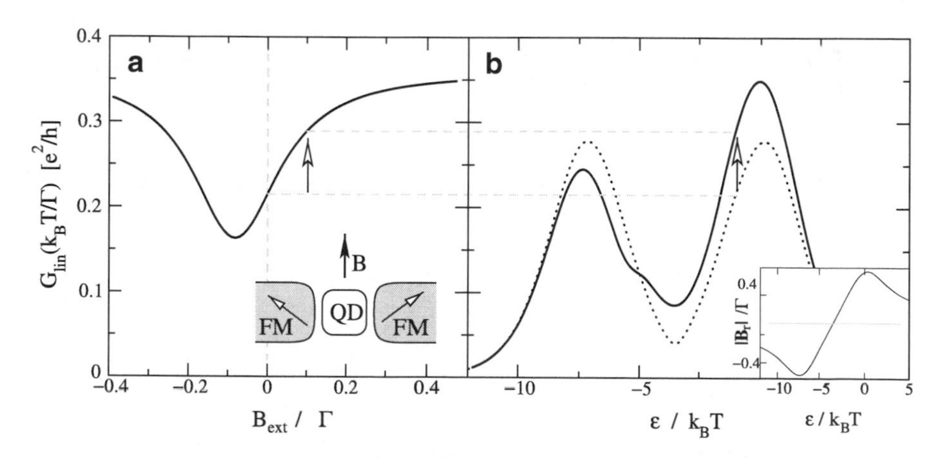

Fig. 6.11 Linear conductance of the dot for an applied external magnetic field $\mathbf{B}_{\text{ext}}$ along $\mathbf{m}_L + \mathbf{m}_R$. **a** Linear conductance as a function of the applied field for $\varepsilon = 0$. **b** Linear conductance as a function of the level position ε without external field (*dotted*) and for the applied field $|\mathbf{B}_{\text{ext}}| = 0.1\Gamma/\hbar$ (*solid*). Further parameters are $\phi = 3\pi/4$, $p = 0.8$, $U = 7k_B T$, and $\tau_{\text{rel}} = 0$. The vertical lines indicate the conductance increase of the dot at $\varepsilon = 0$ due to the applied magnetic field

$p^2\tau_s/\tau_c$ while the width of the dip in Fig. 6.10 provides a direct access to the spin lifetime τ_s. Of course, the conversion of applied magnetic field to frequency requires the knowledge of the Lande factor g, which must be determined separately like in Ref. [27].

6.4.3.2 Magnetic Field Applied Along $\mathrm{m_L} + \mathrm{m_R}$

Finally, we discuss the case of a non-collinear configuration of the leads' magnetizations with a magnetic field applied along the direction $\mathbf{m}_\mathrm{L} + \mathbf{m}_\mathrm{R}$ as shown in Fig. 6.11.

In this case, both the exchange field and the external magnetic field are pointing along the same direction $\mathbf{m}_\mathrm{L} + \mathbf{m}_\mathrm{R}$, so their magnitude is just added. The linear conductance is, then,

$$\frac{G}{G_0} = 1 - p^2 \frac{\tau_\mathrm{s}}{\tau_\mathrm{c}} \frac{\sin^2 \frac{\phi}{2}}{1 + (\mathbf{B}_\mathrm{ext} + \mathbf{B}_\mathrm{L} + \mathbf{B}_\mathrm{R})^2 \tau_\mathrm{s}^2}, \tag{6.26}$$

where ϕ is the angle enclosed by $\mathbf{m}_\mathrm{L}$ and $\mathbf{m}_\mathrm{R}$. The conductance as function of applied magnetic field as plotted in Fig. 6.11a reaches its minimal value when the sum of external and exchange field vanishes. The exchange field leads to a shift of the minimum's position relative to $|\mathbf{B}_\mathrm{ext}| = 0$ [14]. In real experiments, depending on the particular sample geometry, one can expect a magnetic stray field, which is not considered to be part of the experimentally applied magnetic field $\mathbf{B}_\mathrm{ext}$. These stray fields also lead to a shift of the conductance minimum. However, the analyzed setup of external field and magnetizations' directions allows for a stringent experimental verification of spin precession due to the exchange field. To separate the exchange field from the influence of possible stray fields its gate voltage dependence can be used. The exchange interaction as function of the dot gate voltage is plotted in the inset of Fig. 6.11b. While the stray fields does not depend on gate voltage, the exchange field does. In the flat-band limit it even changes sign as a function of gate voltage. By plotting the conductance as function of the gate voltage in Fig. 6.11b, we can observe the typical Coulomb blockade oscillations, when the energy level of the empty or singly-occupied dot becomes resonant with the lead Fermi energy. The interplay of exchange and external field leads to an increase of conductance for one resonance peak, but to a decrease for the other resonance.

6.5 Conclusions

We discussed the possibility to generate, manipulate, and probe single spins in single-level quantum dots coupled to ferromagnetic leads. A finite spin-polarization of the quantum-dot electron is achieved by spin-polarized charge currents from or to the leads at finite bias voltage. Any manipulation of the accumulated spin, e.g., by an external magnetic field or by an intrinsic exchange field, is detectable in the electric current through the device. The occurrence of the exchange field is a consequence of many-body correlations that are one of the intriguing features of nanostructures with large Coulomb interaction.

We determine the dynamics of the quantum-dot spin by deriving expressions for the spin currents through the tunnel barriers. In addition to a contribution that is associated with the spin-polarization of the charge currents from or to ferromagnets, there is a second contribution describing transfer of angular momentum perpendicular to the leads' and dot's magnetization that can be expressed in terms of the exchange field.

In order to manipulate the quantum-dot spin we suggest to make use of the gate- and bias-voltage dependence of the exchange field or to apply an external magnetic field. In particular, the spin precession modifies the dependence of the linear conductance on the opening angle of the lead magnetizations. The degree of modification is tunable by the gate voltage. In nonlinear response, the bias-voltage dependence of the exchange field can give rise to a negative differential conductance. An application of a tunable external magnetic field allows one to determine the dot-spin lifetime and to verify the existence of the intrinsic spin precession caused by the exchange coupling.

Acknowledgments We thank J. Barnaś, G. Bauer, A. Brataas, P. Brower, D. Davidovic, B. Kubala, S. Maekawa, D. Ralph, G. Schön, D. Urban, and B. Wunsch for discussions. This work was supported by the DFG under CFN, SFB 491, and GRK 726, the EC RTN on "Spintronics", Project PBZ/KBN/044/P03/2001, the EC Contract G5MACT-2002-04049, and by the grand for science in years 2006–2008 as a research project.

References

1. Averin, D.V., Likharev, K.K.: In: Altshuler, B.L., Lee, P.A., Webb, R.A. (eds.) Mesoscopic Phenomenon in Solids. Amsterdam, North-Holland (1991)
2. Grabert, H., Devoret, M.H. (eds.) Single Charge Tunneling: Coulomb Blockade Phenomena in Nanostructures. NATO ASI Series B: Physics 294, Plenum Press, New York (1992)
3. Sohn, L.L., Kouwenhoven, L.P., Schön, G. (eds.) Mesoscopic Electron Transport. Kluwer, Dordrecht (1997)
4. Wolf, S.A., Awschalom, D.D., Buhrman, R.A., Daughton, J.M., von Molnár, S., Roukes, M.L., Chtchelkanova, A.Y., Treger, D.M.: Science **294**, 1488–1495 (2001)
5. Zutic, I., Fabian, J., Das Sarma S.: Rev. Mod. Phys. **76**, 323 (2004)
6. Fulton, T.A., Dolan G.J.: Phys. Rev. Lett. **59**, 109 (1987)
7. Baibich, M.N., Broto, J.M., Fert, A., Nguyen Van Dau, F., Petroff, F., Etienne, P., Creuzet, G., Friederich, A., Chazelas J.: Phys. Rev. Lett. **61**, 2472 (1988)
8. Jullière, M.: Phys. Lett. A **54**, 225 (1975)
9. Slonczewski, J.C.: Phys. Rev. B **39**, 6995 (1989)
10. Moodera, J.S., Kinder L.R.: J. Appl. Phys. **79**, 4724 (1996)
11. Jaffrès, H., Lacour, D., Nguyen Van Dau, F., Briatico, J., Petroff, F., Vaurès, A.: Phys. Rev. B **64**, 064427 (2001)
12. M. Braun, J. König, J. Martinek: Europhys. Lett. **72**, 294 (2005)

13. König, J., Martinek, J.: Phys. Rev. Lett. **90**, 166602 (2003)
14. Braun, M., König, J., Martinek, J.: Phys. Rev. B **70**, 195345 (2004)
15. Braig, S., Brouwer, P.W.: Phys. Rev. B **71**, 195324 (2005)
16. Usaj, G., Baranger, H.U.: Phys. Rev. B **63**, 184418 (2001)
17. Usaj, G., Baranger, H.U.: Phys. Rev. B **71**, 179903 (E) (2005)
18. Ono, K., Shimada, H., Ootuka, Y.: J. Phys. Soc. Jpn. **66**, 1261 (1997)
19. Zaffalon, M., van Wees, B.J.: Phys. Rev. Lett. **91**, 186601 (2003)
20. Schelp, L.F., Fert, A., Fettar, F., Holody, P., Lee, S.F., Maurice, J.L., Petroff, F., Vaurés, A.: Phys. Rev. B **56**, R5747 (1997)
21. Yakushiji, K., Mitani, S., Takanashi, K., Takahashi, S., Maekawa, S., Imamura, H., Fujimori, H.: Appl. Phys. Lett. **78**, 515 (2001)
22. Jensen, A., Nygård, J., Borggreen, J.: In: Takayanagi, H., Nitta, J. (eds.) Proceedings of the International Symposium on Mesoscopic Superconductivity and Spintronics, pp. 33–37. World Scientific (2003)
23. Zhao, B., Mönch, I., Vinzelberg, H., Mühl, T., Schneider, C.M.: Appl. Phys. Lett. **80**, 3144 (2002)
24. Tsukagoshi, K., Alphenaar, B.W., Ago, H.: Nature **401**, 572 (1999)
25. Pasupathy, A.N., Bialczak, R.C., Martinek, J., Grose, J.E., Donev, L.A.K., McEuen, P.L., Ralph, D.C.: Science **306**, 86 (2004)
26. Chye, Y., White, M.E., Johnston-Halperin, E., Gerardot, B.D., Awschalom, D.D., Petroff, P.M.: Phys. Rev. B **66**, 201301(R) (2002)
27. Deshmukh, M.M., Ralph, D.C.: Phys. Rev. Lett. **89**, 266803 (2002)
28. Kubetzka, A., Bode, M., Pietzsch, O., Wiesendanger, R.: Phys. Rev. Lett. **88**, 057201 (2002)
29. Sahoo, S., Schönenberger, C., et al.: Nat. Phys. **1**, 102 (2005)
30. Zhang, L.Y., Wang, C.Y., Wei, Y.G., Liu, X.Y., Davidović, D.: Phys. Rev. B **72**, 155445 (2005)
31. Weymann, I., Barnas, J., König, J., Martinek, J., Schön, G.: Phys. Rev. B **72**, 113301 (2005)
32. Weymann, I., König, J., Martinek, J., Barnas, J., Schön, G.: Phys. Rev. B **72**, 115334 (2005)
33. Martinek, J., Utsumi, Y., Imamura, H., Barnaś, J., Maekawa, S., König, J., Schön, G.: Phys. Rev. Lett. **91**, 127203 (2003)
34. Martinek, J., Sindel, M., Borda, L., Barnaś, J., König, J., Schön, G., von Delft, J.: Phys. Rev. Lett. **91**, 247202 (2003)
35. Braun, M., König, J., Martinek, J.: Superlat. Microstruc. **37**, 333 (2005)
36. Meir, Y., Wingreen, N.S.: Phys. Rev. Lett. **68**, 2512 (1992)
37. Brataas, A., Nazarov, Y.V., Bauer, G.E.W.: Eur. Phys. J. B **22**, 99 (2001)
38. Brataas, A., Tserkovnyak, Y., Bauer, G.E.W., Halperin, B.I.: Phys. Rev. B **66**, 060404(R) (2002)
39. König, J., Martinek, J., Barnas, J., Schön, G.: In: Busch, K., et al., (eds.) CFN Lectures on Functional Nanostructures, Lecture Notes in Physics **658**, Springer, 145–164 (2005)
40. Weinmann, D., Häusler, W., Kramer, B.: Phys. Rev. Lett. **74**, 984 (1995)
41. Huettel, A.K., Qin, H., Holleitner, A.W., Blick, R.H., Neumaier, K., Weinmann, D., Eberl, K., Kotthaus, J.P.: Europhys. Lett. **62**, 712 (2003)
42. Ono, K., Austing, D.G., Tokura, Y., Tarucha, S.: Science **297**, 1313 (2002)
43. Johnson, M., Silsbee, R.H.: Phys. Rev. Lett. **55**, 1790 (1985)
44. Johnson, M., Silsbee, R.H.: Phys. Rev. B **37**, 5326 (1988)
45. Epstein, R.J., Fuchs, D.T., Schoenfeld, W.V., Petroff, P.M., Awschalom, D.D.: Appl. Phys. Lett. **78**, 733 (2001)

Chapter 7
Adiabatic Spin Pumping with Quantum Dots

Eduardo R. Mucciolo

7.1 Introduction

In the past few years we have seen important advances in the coherent control of micro- and nano-electronic devices. The experimental effort, driven by the quest for the implementation of quantum computation in semiconductor and superconductor devices, has increased substantially the breath and scope of the study of mesoscopic systems. Progress has been particularly remarkable in lateral semiconductor quantum dots [1]. In these systems, electrons within a two-dimensional gas (2DEG) are confined to small "puddles" by the application of gate voltages. The shape and size of these puddles can be controlled and fine tuned by the same or additional gate voltages. Electrodes also allow one to vary the width of the point contacts that connect the electron puddle to the 2DEG. By acting on these points contacts one can operate the quantum dots in "open" (at least one propagating channel per point contact) or "closed" (fully pinched point contacts) regimes.

A great variety of *stationary* transport phenomena has been observed in these systems over the past 15 years, from discrete Coulomb blockade [1] to signatures of chaotic orbits [2–4] and the Kondo effect [5, 6]. Recently, a new generation of experiments has started to probe the *dynamical* transport properties of quantum dots. A remarkable attempt to explore phase-coherent, pulsed response of an open quantum dot was led by Switkes et al. [7]. Their motivation was the observation of the so-called *adiabatic quantum pumping effect*, first discussed by Thouless more than 20 years ago in the context of one-dimensional electronic systems [8]. Adiabatic quantum pumping takes place when one slowly modulates two or more external parameters of a quantum system, resulting in a net dc current without the

---

E. R. Mucciolo (✉)
Department of Physics, University of Central Florida, P.O. Box 162385, Orlando, FL 32816-2385, USA
e-mail: muccido@physics.ucf.eu

Vojta et al. (eds.), *CFN Lectures on Functional Nanostructures – Volume 2*,
Lecture Notes in Physics 820, DOI: 10.1007/978-3-642-14376-2_7,

application of any bias [9–13]. The effect requires phase-coherent electrons and a system well coupled to reservoirs. Since appearance of Thouless' proposal, quantum pumping has been explained, reinterpreted, discussed, and extended by many authors. Nowadays, a complete and fair review of the literature would take a large portion of this book. Therefore, here we will only refer to those works which are of significance to the main subject of this lecture.

The generation of spin currents in semiconductor heterostructures is a topic of great interest presently [14, 15]. It is believed that spin currents may find applications in in-chip quantum communication, where light propagation is not practical. The full control of the electronic spin degree of freedom in semiconductors also promises to have a large impact in the future of conventional technologies by increasing processing speed, storage capacity, and functionality. While our work does not explore these issues, it does show that quantum dots are versatile enough to yield spin currents with high efficiency, albeit only at low temperatures.

The remaining sections are divided as follows. In Sect. 7.2 we provide a general discussion of adiabatic quantum pumping based on the scattering matrix formalism. In Sect. 7.3 we show that pumping in the presence of a sufficiently strong magnetic field breaks spin symmetry and produces spin-polarized currents. We also estimate the magnitude of the effect under realistic assumptions. The detection of pure spin currents generated with a quantum dot spin pump is discussed in Sect. 7.4, together with an analysis of spurious rectification effects and dephasing. Finally, in Sect. 7.5 we point out to unexplored aspects of quantum pumping and to some promising directions.

7.2 Adiabatic Quantum Pumping

Let us suppose that a certain quantum system is connected to two or more particle reservoirs. Quantum pumping can be defined as the production of net dc currents between reservoirs by acting solely on the quantum system with ac perturbations. No bias is applied between reservoirs. For open quantum systems, when no substantial potential barriers exist between the system and the reservoirs, adiabatic pumping is achieved when the frequency of the ac perturbations is much smaller than the inverse particle dwell time, $\omega \ll 1/\tau_d$. (τ_d is defined as the typical time a particle spends inside the quantum system.) For adiabatic pumping one needs to vary at least two independent parameters, say, X_1 and X_2, to induce a dc current. In Fig. 7.1 we show schematically how a quantum pump can be implemented with a lateral quantum dot.

Is everything quantum in pumping? Actually, no. There are also classical mechanisms that allow one to parametrically drive currents through a system without applying any bias. However, these classical pumps usually require opening and shutting "valves", namely, acting directly on the coupling between the system and the reservoirs. For instance, the human heart pumps blood by a sequence of steps that involve closing and opening valves in coordination with

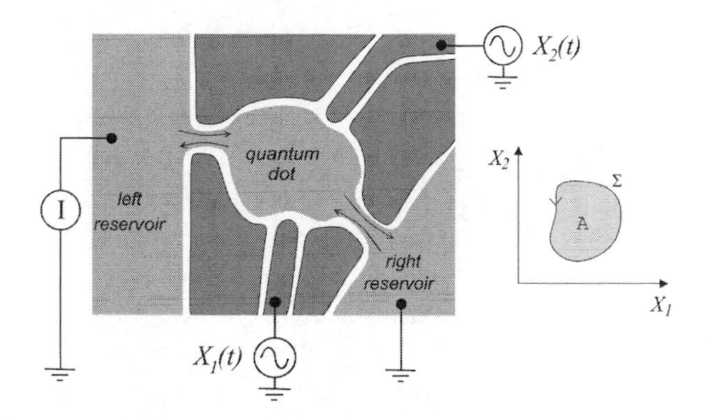

Fig. 7.1 A quantum dot electron pump. The *dark gray* elements represent electrodes and the *arrows* indicate electron flow through the dot-reservoir contact regions. The two gate voltages $X_1(t)$ and $X_2(t)$ act as pumping parameters, continuously deforming the *dot shape*. On the *right-hand side*: a pumping cycle in parameter space. The voltages sweep a closed area $\mathcal{A}$ with contour Σ in parameter space. After a cycle is completed, a net charge is transfered between the two reservoirs

contracting and expanding internal chambers. In a quantum mechanical system, phase coherence allows one to pump matter without acting directly on the system-reservoirs couplings; in that sense, the system is open during the entire pumping cycle. Another difference between classical and quantum pumping is that the former usually requires varying three or more independent parameters (at least two valves and some internal parameter such as potential energy or pressure), while the latter needs only two.

7.2.1 Scattering Matrix Formulation

There is a simple and elegant way to formulate quantum pumping in the language of scattering matrices [11]. Let us define δq_α as the amount of charge that passes through the channel α in one of the contacts during a time interval δt. During this interval, the external parameters X_1 and X_2 vary by $\delta X_1 = \dot{X}_1 dt$ and $\delta X_2 = \dot{X}_2 dt$, respectively. Assuming that δX_1 and δX_2 are small, we can use linear response theory and write

$$\delta q_\alpha(t) = e\left[A_{\alpha;1}\dot{X}_1(t) + A_{\alpha;2}\dot{X}_2(t)\right]\delta t, \tag{7.1}$$

where e is the electron charge and the quantities $A_{\alpha;1,2}$ are called emissitivites. The emissivities depend implicitly on time through $X_1(t)$ and $X_2(t)$. They can be written in terms of the scattering matrix of the quantum system (see below). Notice that (7.1) assumes that the response is instantaneous (local in time). This is correct provided that we are in the weak perturbation, adiabatic regime: $|\dot{X}_j| \ll |X_j|/\tau_d$ for $j = 1, 2$. If we now want to calculate the net charge that flowed through channel α after a full pumping cycle is completed (namely, after a period equal to $2\pi/\omega$), we have that

$$q_\alpha = \int_{\text{cycle}} \delta q_\alpha(t) = e \int_0^{2\pi/\omega} dt \big(\dot{X}_1 A_{\alpha;1} + \dot{X}_2 A_{\alpha;2}\big). \tag{7.2}$$

We can use Green's theorem to convert the integral over time into an area integral in parameter space. The result is

$$q_\alpha = e \int_{\mathcal{A}} dX_1 dX_2 \left(\frac{\partial A_{\alpha;2}}{\partial X_1} - \frac{\partial A_{\alpha;1}}{\partial X_2} \right). \tag{7.3}$$

Notice the similarity between (7.3) and the expression defining a magnetic flux through a loop of area $\mathcal{A}$. Here, the integrand plays the role of a fictitious magnetic field pointing perpendicularly to the $X_1 - X_2$ plane and whose vector potential has in-plane components $A_{\alpha;1}$ and $A_{\alpha;2}$.

The connection to the quantum system scattering matrix $\hat{S}$ occurs through the following expression, first derived by Büttiker et al. [16]:

$$A_{\alpha;j} = \frac{1}{2\pi} \text{Im} \left[\frac{\partial \hat{S}}{\partial X_j} \hat{S}^\dagger \right]_{\alpha\alpha}. \tag{7.4}$$

It is then straightforward to derive the following expression for the total charge pumped through channel α [11]:

$$q_\alpha = \frac{e}{\pi} \int_{\mathcal{A}} dX_1 dX_2 \, \text{Im} \left[\frac{\partial S}{\partial X_2} \frac{\partial S^\dagger}{\partial X_1} \right]_{\alpha\alpha}, \tag{7.5}$$

There are several alternative ways to arrive at (7.5). We would like to refer in particular to the insightful derivation presented by Avrom et al. [17], who also give an interesting physical interpretation of (7.4). The large sensitivity of the scattering matrix elements to interference inside the quantum dot makes the pumped charge a strongly fluctuating quantity.

At this point, the message is that knowing how the quantum dot scattering matrix depends on the external parameters X_1 and X_2 allows one to compute the pumping current in the adiabatic regime. However, there are a few subtle aspects that have been omitted in this discussion. First is the assumption that particles keep their energy as they go in and out of the quantum dot, namely, that the adiabatic process is elastic. This is not exactly true since the system + reservoir Hamiltonian is time-dependent. However, it should be a reasonable approximation in the adiabatic regime, when incoming and outgoing particles see the quantum dot essentially as a nearly static scatterer. Thus, the pumped charge in (7.5) depends on a single, fixed particle energy.

Second, since the reservoirs are Fermi seas, at finite temperatures the particles participating in the transport will have their energies distributed over a continuum. To account for that, we integrate $q_\alpha = q_\alpha(E)$ over energy, appropriately weighted by the derivative of the Fermi–Dirac distribution,

$$Q_\alpha = \int_0^\infty d\varepsilon \left(-\frac{\partial f}{\partial \varepsilon}\right) q_\alpha(\varepsilon)$$
$$= \frac{e}{\pi} \int_0^\infty d\varepsilon \left(-\frac{\partial f}{\partial \varepsilon}\right) \int_{\mathcal{A}} dX_1 dX_2 \, \mathrm{Im}\left[\frac{\partial S}{\partial X_2}\frac{\partial S^\dagger}{\partial X_1}\right]_{\alpha\alpha}, \tag{7.6}$$

with $f(\varepsilon) = 1/[\exp(\varepsilon - \mu)/k_B T + 1]$. Above, it is implicitly understood that $S = S(\varepsilon, X_1, X_2)$ and that the two reservoirs have the same equilibrium properties, namely temperature T and chemical potential μ (i.e., no bias is applied). It is clear that (7.6) is reduced to (7.5) evaluated at the Fermi energy as $T \to 0$.

At this point it is worthwhile deriving an expression for the dc component of the pumping current in some simple situation. For instance, let us assume that the pumping parameters vary as $X_1(t) = X_{01} + \delta X_1 \cos(\omega t)$ and $X_2(t) = X_{02} + \delta X_2 \cos(\omega t - \varphi)$, with amplitudes δX_1 and δX_2 so small that the integrand in (7.6) is essentially constant over the pumping cycle. It is not difficult to show that, in this case,

$$I_\alpha = \frac{\omega Q_\alpha}{2\pi} \approx e\omega \sin\varphi \; \delta X_1 \delta X_2 \mathcal{I}_\alpha, \tag{7.7}$$

where

$$\mathcal{I}_\alpha = \frac{1}{\pi} \int_0^\infty d\varepsilon \left(-\frac{\partial f}{\partial \varepsilon}\right) \mathrm{Im}\left[\frac{\partial S}{\partial X_2}\frac{\partial S^\dagger}{\partial X_1}\right]_{\alpha\alpha}. \tag{7.8}$$

In this regime, the pumping current is a bilinear function of the pumping amplitudes and has a sinusoidal dependence on the phase difference ϕ.

The small-amplitude approximation ceases to be valid when $|X_i| \gg X_i^{(c)}$, $i = 1, 2$, where $X_1^{(c)}$ and $X_2^{(c)}$ are the characteristic parameter scales over which the scattering matrix elements change substantially. For large parameter amplitudes, the integrand in (7.6) will fluctuate and change sign many times within the integration area $\mathcal{A}$. As a result, the pumping charge will depend on the parameter amplitudes as $\sqrt{\delta X_1 \delta X_2}$ rather than bilinearly [11]. Moreover, one expects to observe a more complicated dependence on ϕ.

Most of these features were observed in a experiment by Switkes et al. [7] using an open lateral quantum dot subjected to two ac, shape deforming gate voltages. In that experiment, the electron dwell time in the quantum dot was estimated to be under 1 ns, while the pumping frequency used was in the tens of MHz. Therefore, the pumping regime was certainly adiabatic. However, a few aspects of the experimental data were in conflict with the theoretical predictions [11, 13]. The discrepancies involved the amplitude of the quantum pumping current and its symmetry properties in the presence of an external magnetic field. In the next section we briefly describe the statistics and symmetry of pumping currents. As pointed out by Brouwer [18], that early experiment was likely plagued by spurious rectification effects due the capacitive coupling between gates and electron reservoirs.

7.2.2 Statistics and Symmetry Properties of Pumping Current

Phase coherence is at the heart of adiabatic quantum pumping. The pumping current amplitude is a very sensitive function of the spatial structure of the wave function inside the quantum dot. That structure in turn is a manifestation of electron interference and therefore very sensitive to changes in the shape of the confining potential. One expects that different realizations of the dot geometry will lead to marked changes in the amplitude and direction of the pumping current. A weak magnetic field that scrambles the electron phase with little effect on the orbital motion will also cause a similar effect. Such mesoscopic fluctuations were indeed observed in Ref. [7].

There is a vast literature treating the statistical properties of the scattering matrix in chaotic mesoscopic systems (see Ref. [19] for a review). Using this accumulated knowledge, it is possible to determine the full distribution of the pumping current for all relevant universal symmetry classes and for any number of propagating channels in the contacts [11]. At zero temperature, it was found that the distribution ranges from nearly Gaussian when the contacts carry many propagating channels ($N = N_R + N_L \gg 1$) to a singular form with power-law tails when $N_R = N_L = 1$ (one propagating channel per contact). For the single-channel case, $P(I) \sim |I|^{-\kappa}$ for large I, where $\kappa = 9/4$ (3) when time-reversal symmetry is present (absent). Thus, while the ensemble average of the pumping current is always zero at zero bias, $\langle I \rangle = 0$, the variance is always nonzero and actually diverges for the $N_R = N_L = 1$ case. In practice, these very large mesoscopic fluctuations are capped by the electron's finite decoherence time [20, 21]. For $N > 2$, thermal fluctuations also decrease the amplitude of the pumping current.

The effect of discrete spatial symmetries and magnetic field inversion on the adiabatic quantum pumping current can also be understood through the scattering matrix formalism described in Sect. 7.2.1. After some initial controversy in the literature, this issue was lucidly discussed in Ref. [22]. The most important property to mention in regard to symmetry is the following: For open quantum dots with no discrete spatial symmetry, there is no particular symmetry in the pumping current with respect to the inversion of the magnetic field:

$$I(B) \neq I(-B) \quad \text{for a completely asymmetric dot.} \tag{7.9}$$

In other words, there is no counterpart to the Onsager relation characteristic of biased dissipative transport. The experiment of Ref. [7] was not consistent with this prediction. This fact and others indicate that this early attempt to observe quantum pumping was likely dominated by rectification effects (which should yield $I(B) = I(-B)$).

Interestingly, it is also possible to show that for a pump with inversion symmetry (left-right and top-bottom symmetries), there is always a perfect cancellation of charge flow during the pumping cycle, yielding $I = 0$ identically.

7.3 Pumping Spin with Quantum Dots

The quantum pumping current described in Sect. 7.2 contained spin-degenerate electrons. Therefore, the contributions coming from "up" and "down" spin components of the charge current had identical direction and amplitude, leading to zero net spin or angular momentum transport. In order to generate a net spin flow out of spin-depolarized electron sources, one needs to add a spin-symmetry breaking field to the system. There are two simple ways to do that: (1) creating a Zeeman splitting in the spectrum by applying an external magnetic field, or (2) using a pump with strong spin–orbit coupling. While the latter has been recently studied theoretically in different contexts [23, 24], it remains very challenging to implement experimentally. Here we will focus on the former case [25], which has already been tested and shown to produce clear evidence of spin-polarized transport [26].[1]

The main idea can be understood through the scheme shown in Fig. 7.2. Since the materials underlying reservoirs and quantum dot is the same, upon the application of an external magnetic field B, energy levels inside and outside the dot will be spin split by the Zeeman energy $E_Z = g^* \mu_B B/2$, where g^* is the effective gyromagnetic factor and μ_B is the Bohr magneton. Let us assume that E_Z is much smaller than the Fermi energy yet sizeable in comparison to the mean level spacing inside the dot, Δ. The Zeeman splitting in the reservoirs amounts to a small shift in the wavelengths of the "up" and "down" electron states at the Fermi surface and nothing else. However, the effect in the quantum dot states can be much more pronounced. Since there are marked differences in the spatial distribution of eigenfunctions of states even if they are close in energy, changing the orbital content of states near the Fermi level will strongly affect the pumping current. Recall that the matrix elements of the scattering matrix fluctuate in energy for systems which have a chaotic dynamics in the classical limit. Therefore, by having $E_Z > \Delta$ we make the "up" and "down" components of the pumping current close to uncorrelated for a chaotic pump. If we define charge and spin pumping currents as[2]

$$I_c = I_\uparrow + I_\downarrow \tag{7.10}$$

and

---

[1] Spin pumping in interacting nanowires was first discussed in Ref. [27] and further extended in Ref. [28].

[2] Here for convenience, we have adopted the same units for charge and spin currents.

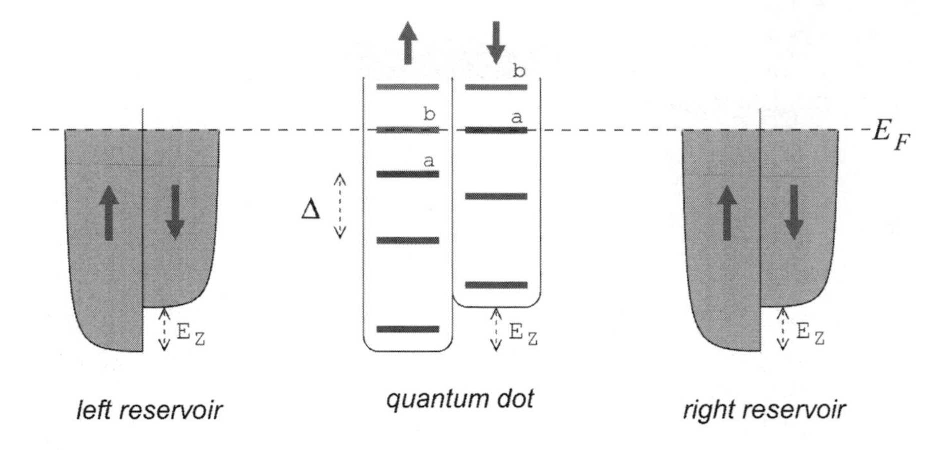

Fig. 7.2 Schematic view of the states involved in the quantum pumping current through a dot in the presence of a Zeeman energy splitting E_Z of the order of the mean level spacing Δ. The *solid arrows* indicate spin orientation. The Zeeman splitting does not significantly modify the nature of the states in the reservoirs found near the Fermi level. However, if *a* and *b* are states with distinct wave function content, the splitting can make $I_\uparrow \neq I_\downarrow$

$$I_s = I_\uparrow - I_\downarrow, \tag{7.11}$$

respectively, we see that while $I_s(B = 0) \equiv 0$ due to spin degeneracy, typically we have $I_s(B > B_c) \neq 0$, with B_c being a characteristic Zeeman field related to the pumping current correlation energy, E_c: $B_c = E_c/g^*\mu_B$. For weakly coupled quantum dots at very low temperatures E_c is equal to Δ. However, energy levels in open dots are broad resonances instead of sharp discrete levels. Moreover, lowest achievable temperatures with present technologies are comparable to the level spacing found in all but the smallest quantum dots. Thus, in the general case, $E_c = \max\{k_B T, \Delta, \hbar\gamma\}$, where γ is the electron escape rate (i.e., the inverse dwell time). This scale is not too large: for a ballistic GaAs quantum dot[3] with a 1-μm in linear size and one open propagating channel per lead, one usually finds $\Delta = \pi\hbar\gamma \approx 10\mu\text{eV}$, leading to $B_c \approx 1$ T at temperatures below 100 mK.

In practice, one should avoid using a magnetic field perpendicular to the 2DEG underlying a lateral quantum dot. This is because even at fields of about 1 T there is already a significant reduction on the sensitivity of the wave functions to external perturbations (such as shape distortions) due to the formation of Landau states. This in turn reduces the dependence of the scattering matrix elements on parametric driving and, consequently, decreases the pumping current amplitude. Thus, a parallel magnetic field that only couples significantly to the electron spin and leaves the orbital motion unaltered is a more sensible choice for producing spin-polarized currents. The drawback is that this choice limits the spin polarization to only one direction.

---

[3] Electron densities are usually around 10^{11} cm^{-2} in high-quality GaAs wafers.

7.3.1 Pure Spin Currents

We have argued that phase coherence combined with wave function sensitivity to parametric changes make "up" and "down" spin components of an adiabatic pumping current nearly independent when even a moderate magnetic field is applied. This effect can be explored to produce dc spin transport with zero net charge transfer, the so-called pure spin current. The mechanism is illustrated in Fig. 7.3. The idea here is again based on the large sensitivity of confined quantum states to parametric changes when the underlying electronic motion is classically chaotic [29]. If a third tuning parameter, X_3 is provided besides the other two used to drive the system adiabatically, X_1 and X_2, such that $H = H(X_1, X_2, X_3)$, one can try to search for a realization of the Hamiltonian when "up" and "down" spin components of the pumping current have the same amplitude but opposite directions. This point is denoted in Fig. 7.3 by X_{ps}. Notice that $I_s \neq 0$ while $I_c = 0$ at this point.

7.3.2 Quantitative Analysis

When no spin-symmetric breaking field exists, both "up" and "down" spin components of the pumping current are identical. For an irregularly shaped

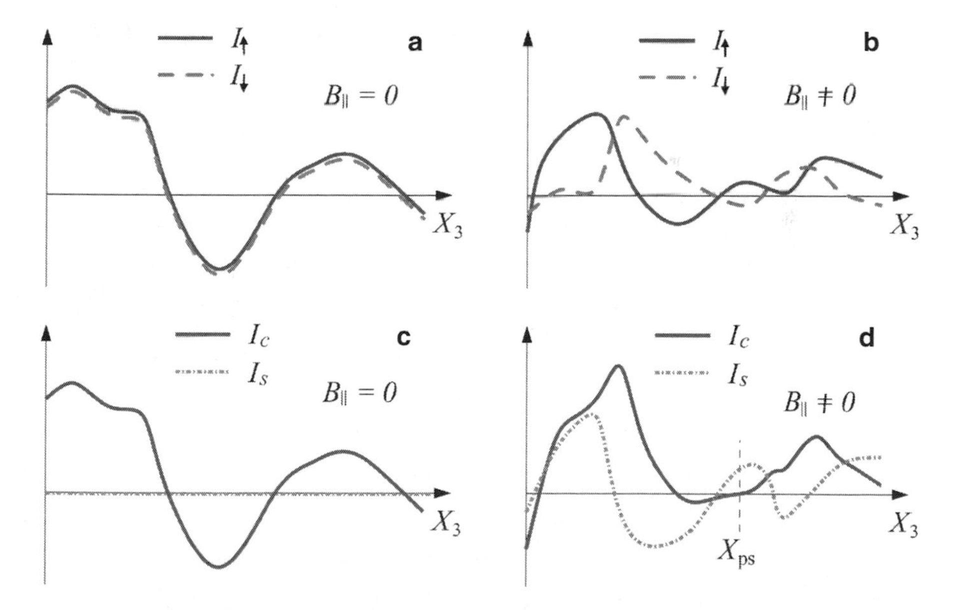

Fig. 7.3 Schematic plot of dc pumping currents as functions of a quantum dot tuning parameter X. The plots in (**a**) and (**b**) show the spin up and spin down components of the pumping current, $I_{\uparrow,\downarrow}$, the total charge I_c, and total spin currents I_s when no parallel magnetic field is applied, $B_\parallel = 0$. Notice that in this case $I_s = 0$. In plots (**c**) and (**d**) $B_\parallel \neq 0$, making $I_\uparrow \neq I_\downarrow$ and $I_s \neq 0$, in general. There are values of the tuning parameter, such as X_{ps}, where a finite spin current occurs without net charge transport ($I_c = 0$)

quantum dot, as the amplitude of the external magnetic field is increased past the characteristic field B_c, the two spin components become uncorrelated. Since the ensemble averaged value of the dc pumping current is zero in the absence of bias, $\langle I_\uparrow \rangle = \langle I_\downarrow \rangle = 0$, we can write that

$$\mathrm{corr}_{\uparrow\downarrow}(B) \equiv \langle I_\uparrow I_\downarrow \rangle = \begin{cases} \langle I_\uparrow^2 \rangle, & B = 0, \\ 0, & B \gg B_c. \end{cases} \tag{7.12}$$

For intermediate values of the magnetic field, the correlation interpolates monotonically between the two limiting values.

The full spin polarization of the pumping current only occurs at those special configurations where $I_\uparrow = I_\downarrow$ exactly, regardless to how large the magnetic field is (provided it is nonzero). For all other configurations, the polarization will be smaller, random, and dependent on the magnetic field. In order to quantify the typical amplitude of the spin current in comparison to the charge current, we introduce the quantity $r_{\mathrm{pol}} = \sqrt{\langle I_s^2 \rangle / \langle I_c^2 \rangle}$. Both charge and spin currents can be written in terms of the correlation function $\mathrm{corr}_{\uparrow\downarrow(B)}$. It is then straightforward to show that

$$r_{\mathrm{pol}} = \sqrt{\frac{\mathrm{corr}_{\uparrow\downarrow}(0) - \mathrm{corr}_{\uparrow\downarrow}(\mathrm{B})}{\mathrm{corr}_{\uparrow\downarrow}(0) + \mathrm{corr}_{\uparrow\downarrow}(B)}} = \begin{cases} 0, & B = 0, \\ 1, & B \gg B_c. \end{cases} \tag{7.13}$$

This means that a substantial spin polarization may be achieved for sufficiently large magnetic fields, as we have anticipated using qualitative arguments in Sect. 7.3.1. Therefore, it is important to be able to quantify this effect, as well as the characteristic dependence of the spin current amplitude on temperature and on the number of propagating channel in the leads.

The non-trivial aspect of evaluating $\mathrm{corr}_{\uparrow\downarrow}(B)$ is that one needs to find a formulation where information about the energetics inside the dot can be incorporated. In other words, one has to be able to track down how wave functions and energy levels depend not only on the driving parameters but also on how scattering matrix elements corresponding to different spin orientations become progressively uncorrelated as the Zeeman splitting widens. Thus, a microscopic Hamiltonian formulation is unavoidable and the calculation cannot be performed within the static, random scattering matrix approach used by Brouwer [11].

Luckily, however, there are at least two ways to carry out the analytical calculation of $\mathrm{corr}_{\uparrow\downarrow}(B)$. Moreover, it is also possible to complement the analytical calculations with numerical simulations when the former become too involved. In our original work [25], we opted for suitably adapting a formulation developed by Vavilov and coworkers for the spinless *charge* pumping [30] variance. Their approach used the non-equilibrium Keldysh technique to relate the instantaneous pumping current to the *real-time* scattering matrix of the dot. Being very general and applicable beyond the adiabatic approximation, this formulation is rather sophisticated. However, as we show below, the extension to the non-zero Zeeman field case for the adiabatic regime is simple, can be worked out from the expressions found in Ref. [30], and yields the information about the correlation between the spin components of the pumping current that we are seeking.

Before going into the details of the calculation, it is important to highlight the assumptions used in the model. Vavilov and coworkers computed the pumping current assuming that: (1) the electron eigenstates of a non-interacting dot could be described by the unitary ensemble of Gaussian random matrix and (2) that the leads contained many propagating channels. These assumptions are well justified for the case of open dots with sufficiently complicated geometry, large contacts, and in the presence of a time-reversal symmetric breaking field. In practice, the latter condition can be enforced by letting the external magnetic field have a weak (tens of mT) perpendicular component.

Following Ref. [30], let us divide the total Hamiltonian of the total system into three parts,

$$H_{\text{total}} = H_{\text{dot}} + H_{\text{leads}} + H_{\text{dot-leads}}. \tag{7.14}$$

For the dot Hamiltonian we have

$$H_{\text{dot}} = \sum_{n,m=1}^{M} \sum_{\sigma=\pm 1} \left(\mathcal{H}_{n,m} + \delta_{n,m} \frac{\sigma E_Z}{2} \right) a_{n\sigma}^{\dagger} a_{m\sigma}, \tag{7.15}$$

where $a_{n\sigma}^{\dagger}, a_{n\sigma}$ are creation and annihilation electron operators defined over a single-particle basis of size M and $\{\mathcal{H}_{n,m}\}$ denote matrix elements of the orbital contribution to the electron energy on that basis. The spin-dependent, diagonal term accounts for the Zeeman energy. Notice that $\mathcal{H} = \mathcal{H}(t)$. For the leads Hamiltonian we have

$$H_{\text{leads}} = \sum_{\alpha=1}^{N} \sum_{k} \sum_{\sigma=\pm 1} \left[E(k) + \frac{\sigma E_Z}{2} \right] c_{\alpha k\sigma}^{\dagger} c_{\alpha k\sigma}, \tag{7.16}$$

where $c_{\alpha k\sigma}^{\dagger}, c_{\alpha k\sigma}$ are creation and annihilation electron operators in the leads, the index α runs over all $N = N_R + N_L$ propagating channels in the right and left leads, and $E_\alpha(k)$ is the electron energy dispersion relation in the channel α. Finally, for the dot-leads coupling Hamiltonian, we have

$$H_{\text{dot-leads}} = \sum_{\alpha,k,n,\sigma} \left(W_{n\alpha}\, c_{\alpha k\sigma}^{\dagger}\, a_{n\sigma} + \text{h.c.} \right), \tag{7.17}$$

where $W_{n\alpha}$ are coupling matrix elements related to the overlap between dot and lead single-particle wave functions at the contact regions. These matrix elements are assumed energy and spin independent.[4]

---

[4] It is not too restrictive to assume that $W_{n\alpha}$ does not depend on the incoming or outgoing particle energy. The pumping current will still depend on the chemical potential in the leads through the Fermi distribution functions and the fact that $W_{n\alpha}$ depends on the dot state through n.

The dot single-particle Hamiltonian is separated into static and time-dependent or driving terms,

$$\mathcal{H}(t) = \mathcal{H}_0 + \lambda_1(t)X_1 + \lambda_2(t)X_2. \tag{7.18}$$

The static or unperturbed term $\mathcal{H}_0$ is taken to be a member of the Gaussian unitary ensemble with variance $\left\langle |[\mathcal{H}_0]_{n,m}|^2 \right\rangle = M\Delta^2/\pi^2$, where $M \gg 1$ is the matrix rank and Δ is the level space near the band center and where the Fermi level is located. The scalar functions $\lambda_1(t)$ and $\lambda_2(t)$ modulate the amplitude of the time-dependent perturbation. One usually chooses phase-locked, harmonic functions: $\lambda_1(t) = \cos(\omega t)$ and $\lambda_2(t) = \cos(\omega t + \varphi)$. It turns out that for sufficiently large dots, the diagonal parts of X_1 and X_2 are strongly suppressed and these matrices can be taken traceless. The strength of the perturbation is then fully characterized by only three numbers, namely Tr (X_1^2), Tr (X_2^2), and Tr $(X_1 X_2)$. These quantities can be related to the so-called "velocity" correlator [31], which measures how energy levels in the quantum dot respond to a linear, static perturbation of the form $\lambda_1 X_1 + \lambda_2 X_2$:

$$\frac{2\mathrm{Tr}(\mathrm{X_i X_j})}{M^2} = \left\langle \frac{\partial \varepsilon_a}{\partial \lambda_i} \frac{\partial \varepsilon_a}{\partial \lambda_j} \right\rangle - \left\langle \frac{\partial \varepsilon_a}{\partial \lambda_i} \right\rangle \left\langle \frac{\partial \varepsilon_a}{\partial \lambda_j} \right\rangle, \tag{7.19}$$

where $\{\varepsilon_a(\lambda_1, \lambda_2)\}$ are the energy eigenvalues of the *isolated* quantum dot. This velocity correlator can be measured, thus providing information about the traces.

Two fundamental relations have to be used in this approach to write the pumping current in terms of the dot Hamiltonian. The first one is the standard connection between the scattering matrix and the scatterer (dot) Hamiltonian,

$$S_{\alpha\beta;\sigma}(t,t') = \delta_{\alpha,\beta}\delta(t-t') - 2\pi i \nu \sum_{n,m} W^*_{n\alpha} G^{(R)}_{nm;\sigma}(t,t') W_{m\beta}, \tag{7.20}$$

where ν is the density of states at the Fermi level and $G^{(R)}_{nm;\sigma}(t, t')$ is the retarded Green's function of the *open* dot, satisfying the matrix equation

$$\left[i\hbar\frac{\partial}{\partial t} - \mathcal{H}(t) - \frac{\sigma E_Z}{2} + i\pi\nu W W^\dagger \right] G^{(R)}_\sigma(t,t') = \delta(t-t'). \tag{7.21}$$

The second relation is a connection between the total current flowing into one of the reservoirs[5] and the scattering matrix in real time,

$$I_\sigma(t) = e\sum_\alpha \Lambda_{\alpha\alpha} \int\int dt_1 dt_2 \left[\sum_\beta S_{\alpha\beta;\sigma}(t,t_1) \tilde{f}_\beta(t_1-t_2) S^*_{\alpha\beta;\sigma}(t_2,t) - \tilde{f}_\alpha(+i0) \right], \tag{7.22}$$

where $\tilde{f}_\alpha(t)$ is the inverse Fourier transform of the electron Fermi–Dirac distribution function in the α channel,

[5] Since charge is not accumulated during the pumping cycle, all current that flows from one of the reservoirs enters the other: $I^{\mathrm{left}} = -I^{\mathrm{right}}$

$$\tilde{f}_\alpha(t) = \int_{-\infty}^{\infty} \frac{d\varepsilon}{2\pi} e^{i\varepsilon t/\hbar} \left[\frac{1}{e^{(\varepsilon-\mu_\alpha)/k_BT}+1} - \frac{1}{2} \right] = \frac{ik_BTe^{i\mu_\alpha t/\hbar}}{2\sinh(\pi k_BTt/\hbar)} \tag{7.23}$$

and Λ is the auxiliary matrix

$$\Lambda_{\alpha\beta} = \delta_{\alpha,\beta} \begin{cases} N_R/N, \alpha \in \text{right lead}, \\ -N_L/N, \alpha \in \text{left lead}. \end{cases} \tag{7.24}$$

Equation (7.22) was derived in Ref. [30] under the standard assumption that the dispersion relation in the leads can be linearized: $E(k) = v_F k$, where $v_F = 1/2\pi\nu$ is the Fermi velocity. Inserting (7.20) into (7.22) and assuming that no bias is present (such that we can drop the index α in both μ_α and f_α), it was found that

$$I_\sigma(t) = e \int\int dt_1 dt_2 \tilde{f}(t_1 - t_2) \mathrm{Tr}[R_\sigma(t, t_1, t_2)], \tag{7.25}$$

where we have introduced the matrix

$$R_\sigma(t, t_1, t_2) = 2\pi i\nu\, W^\dagger G_\sigma^{(R)}(t, t_1)[\mathcal{H}(t_1) - \mathcal{H}(t_2)]\, G_\sigma^{(A)}(t_2, t) W\Lambda. \tag{7.26}$$

At this point, there is an trick that can used when dealing with spin currents: Since the Zeeman energy splitting is uniformly present (inside and outside the quantum dot), it can be accounted for by a spin-dependent shift of the chemical potential. More specifically, if we make the substitution

$$\mu \to \mu + \frac{\sigma E_Z}{2}, \tag{7.27}$$

we can easily see that

$$\tilde{f}(t) \to e^{i\sigma E_Z t/2} \tilde{f}(t). \tag{7.28}$$

We can then drop the spin index from the Green's functions in (7.26) and set the Zeeman energy to zero inside the dot. This allows us to reduce the ensemble-averaged correlator of dc spin components of the pumping current to

$$\begin{aligned} \mathrm{corr}_{\uparrow\downarrow}(B) &= \left(\frac{\omega}{2\pi}\right)^2 \int_0^{2\pi/\omega} dt \int_0^{2\pi/\omega} dt' \langle I_\uparrow(t) I_\downarrow(t')\rangle \\ &= \left(\frac{e\omega}{2\pi}\right)^2 \int\int\int\int dt_1\, dt_2\, dt_1'\, dt_2'\, e^{iE_Z(t_1 - t_2 + t_1' - t_2')/2} \\ &\times \int_0^{2\pi/\omega} dt \int_0^{2\pi/\omega} dt' \left\langle \mathrm{Tr}[R(t, t_1, t_2)] \mathrm{Tr}\left[R(t', t_1', t_2')\right] \right\rangle. \end{aligned} \tag{7.29}$$

Thus, except for the Zeeman energy-dependent exponential factor in the integrand, the calculation amounts to same one performed by Vavilov and coworkers. They used a diagramatic technique based on random-matrix theory to evaluate the correlator $\langle \mathrm{Tr}[R(t,t_1,t_2)]\mathrm{Tr}\left[R(t',t'_1,t'_2)\right]\rangle$ in the limit of $N \gg 1$ and for perfectly transparent leads. The dimensionless parameter $1/N$ was used to regroup diagrams to leading order, yielding two-particle diffusion propagators similar to those used in the theory of disordered mesoscopic systems. The steps involved in their derivation are quite lengthy and we encourage the reader to check Ref. [30] for the details.

We can take their expression for $\langle \mathrm{Tr}[R(t,t_1,t_2)]\mathrm{Tr}\left[R(t',t'_1,t'_2)\right]\rangle$ [see (7.29) in Ref. [30]] and add the Zeeman-energy dependent exponential factor into the time integrations present in (7.29). In the adiabatic bilinear regime, the final expression for the spin current correlator becomes relatively compact:

$$\mathrm{corr}_{\uparrow\downarrow}(B) = \frac{e^2\omega^2 g\tau_\mathrm{d}\det(C)\sin^2\varphi}{2\pi^2}\int_0^\infty d\tau e^{-\tau/\tau_\mathrm{d}}\left(1+\frac{\tau}{\tau_\mathrm{d}}\right) \times \left[\frac{k_B T\tau/\hbar}{\sinh(\pi k_B T\tau/\hbar)}\right]^2 \cos(\tau E_Z/2\hbar), \tag{7.30}$$

where $g = N_R N_L/N$ is the dot dimensionless conductance, $\tau_\mathrm{d} = 2\pi\hbar/N\Delta$ is the dwell time, and

$$C_{ij} = \frac{\pi}{M^2\Delta}\mathrm{Tr}\left(X_i X_j\right). \tag{7.31}$$

It is important to remark that (7.30) can also be derived using a semiclassical, trace formula representation of the scattering matrix. In that case, the validity of the semiclassical formulation is guaranteed by the large number of channels in the leads and the presumed chaotic electronic motion inside the dot. Details of the semiclassical calculation can be found in Ref. [21].

In Fig. 7.4 we show the resulting r_pol as a function of magnetic field for different temperatures and escape rates. Notice that thermal fluctuations have a relatively stronger effect on the polarization than the energy spreading due to the finite electron dwell time in the dot. The curves for the smallest values of N should be taken just as a qualitative indication of the dependence since (7.30) is only valid in the large-N limit.

7.4 Spin Current Detection

Once a spin-polarized current is generated through the mechanism presented in Sect. 7.3, the question that naturally arises is how to detect or measure the spin polarization. One could imagine using ferromagnetic leads to spin filter the current, like polarizers are used to filter out any component of a light beam. However,

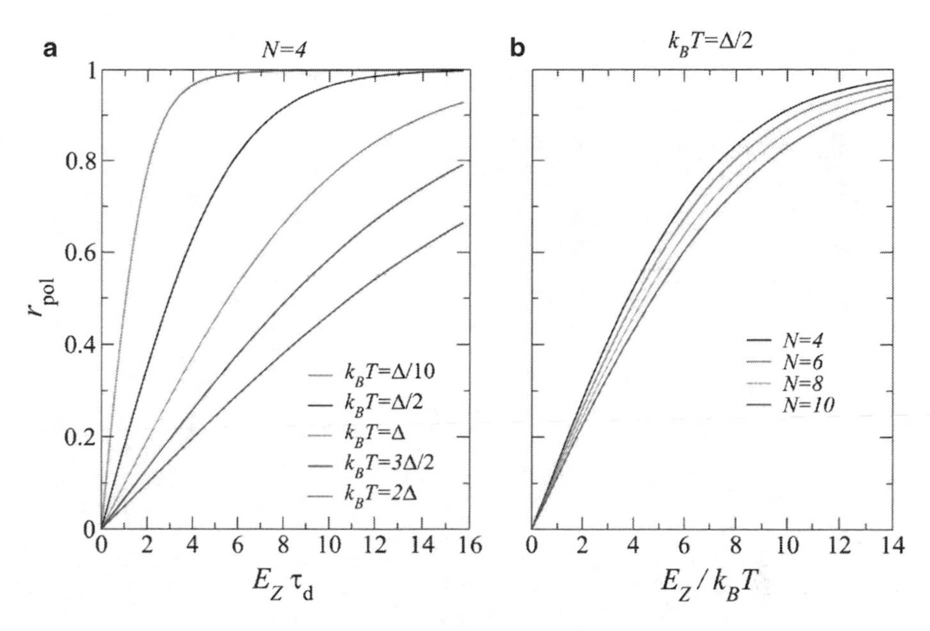

Fig. 7.4 The dependence of the relative spin polarization on the Zeeman splitting energy for (**a**) different temperatures and (**b**) for different number of propagating channels in the leads. Notice the different scales in the horizontal axes

adding ferromagnetic leads is not yet a realistic option for lateral quantum dot setups. A more appropriate and readily available option is to use quantum point contacts (QPCs) where the width of the constriction can be controlled by gate voltages [32]. A schematic view of such device connected to a spin pump is shown in Fig. 7.5.

The detection works as follows. At sufficiently low temperatures and in the absence of a spin-splitting magnetic field, the conductance in the QPC is spin degenerate and presents steps at even values of the conductance quantum e^2/h [33]. When the spin-splitting field is turned on, the degeneracy is broken and intermediate steps appear at odd values of e^2/h. In this case, if one operates the QPC in the range where just one propagating channel is allowed, only the "up" spin component of the current will be transmitted and the "down" component will be filtered out. When the quantum pump is set to work at a configuration where $I_\uparrow = -I_\downarrow$ as to have a zero net flow of charge, the QPC will block $I_\downarrow$, resulting in $I_c \neq 0$ through the constriction and a nonzero voltage drop across it ($V_{\text{qpc}} \neq 0$). However, if the QPC conductance is brought to the second plateau at $2e^2/h$, both spin components are allowed to flow, making $I_c = 0$ and $V_{\text{qpc}} = 0$ as well. Thus, by monitoring the voltage across the QPC the number of propagating channels goes from odd to even, it is possible to detect the spin current. When the pump works away from the pure spin configuration, $I_\uparrow$ and $I_\downarrow$ do not fully compensate each other and the charge current never drops to zero when the number of propagating channels in the QPC is even.

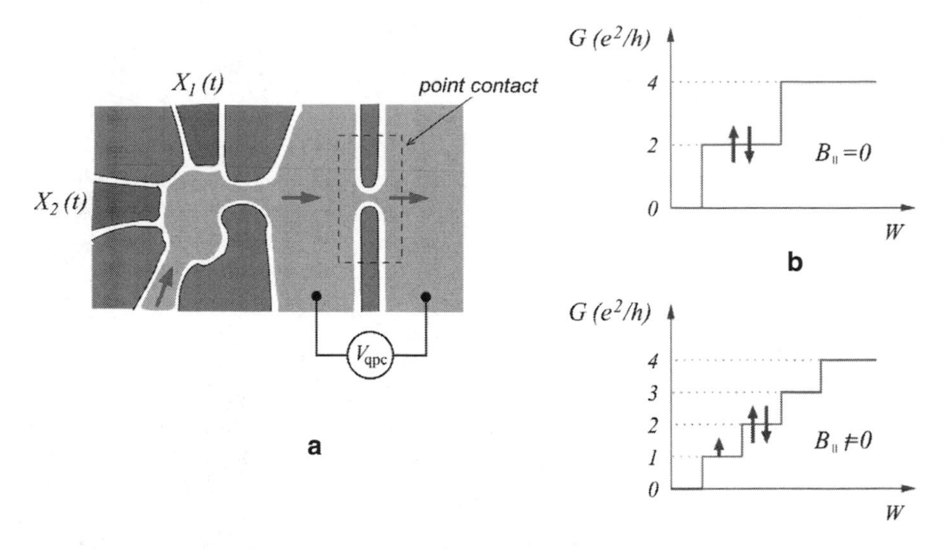

Fig. 7.5 **a** Schematic illustration of a quantum dot spin pump connected to a spin filtering point contact. The *arrows* indicate the direction of the current. **b** Linear conductance of the point contact as a function of the constriction width in the absence and in the presence of a parallel magnetic field

This scheme was implemented experimentally by Watson et al. [26]. In their setup a moderate perpendicular magnetic field was used to focus the current flow into the QPC (the cyclotron radius was considerably larger than the linear size of the effective well inside the dot, thus Landau level quantization was not sufficiently strong to impair parametric pumping). Their results show clear evidence that spin polarization is achieved by parametrically driving the quantum dot in the presence of a parallel magnetic field. The spin current observed corresponded to tens of $\hbar$ per cycle at a frequency of 10 MHz. While this seems to confirm at least qualitatively our proposal, a complete test would also require collecting enough statistics to compare the statistics of the spin current to the theoretical predictions based on random matrix theory, as discussed in Sect. 7.4. This has yet to be done.

7.4.1 The Effect of Rectification

Giving the dominant presence of rectification effects in the earlier quantum pumping experiment by Switkes et al. [7], one could expect that a similar situation occurred in the experiment described in Ref. [26]. Indeed, this possibility cannot be discarded without a careful symmetry and statistical analysis of the pumping current. However, it turns out that the generation of spin-polarized currents is somewhat insensitive to the predominant mechanism, namely, even rectification due to spurious capacitive coupling between gate electrodes and the 2DEG would still lead to spin polarization. This can be easily understood if we recall that the conductance in the dot is susceptible to mesoscopic fluctuations during the pumping cycle. Therefore, dot conductance will vary in time as the pumping parameters run over a cycle. Now, if a certain spurious bias voltage $\delta V(t)$ occurs during the pumping cycle, an instantaneous rectification current

$$\delta I_{\uparrow,\downarrow}(t) = G_{\uparrow,\downarrow}(t)\delta V(t) \tag{7.32}$$

will be added to the quantum pumping current. When the parallel magnetic field is present, $G_{\uparrow}(t) \neq G_{\downarrow}(t)$ since each spin component of the current is carried through the dot by a different composition of electronic states. As a result, after averaging over a full cycle, we will find $\overline{\delta I_{\uparrow}(t)} \neq \overline{\delta I_{\downarrow}(t)}$, leading to an additional contribution to the dc spin current.

A quantitative study of rectification effects requires a model circuit where the capacitive coupling between electrodes and leads is incorporated in a realistic way. Following the circuit model suggested by Brouwer [18], we have attempted to characterize the statistical properties of spin currents generated by rectification in Ref. [21]. The model circuit is shown in Fig. 7.6 for the case of a current setup. It is straightforward to find that [18, 21]

$$\delta I_{\uparrow,\downarrow}(t) = R_A G_{\uparrow,\downarrow}\left(C_{2L}\frac{dX_1}{dt} + C_{1L}\frac{dX_2}{dt}\right). \tag{7.33}$$

Using the semiclassical method, the correlation between "up" and "down" spin components of the rectified current was calculated under the same assumptions that led to (7.30). We have found that

$$\begin{aligned} \mathrm{corr}^{\mathrm{rect}}_{\uparrow\downarrow}(B) &= \langle \delta I_{\uparrow}\delta I_{\downarrow}\rangle \\ &= \mathcal{C}\int\limits_0^{\infty} d\tau e^{-\tau/\tau_d}\left[\frac{k_B T\tau/\hbar}{\sinh(\pi k_B T\tau/\hbar)}\right]^2 \cos(\tau E_Z/2\hbar), \end{aligned} \tag{7.34}$$

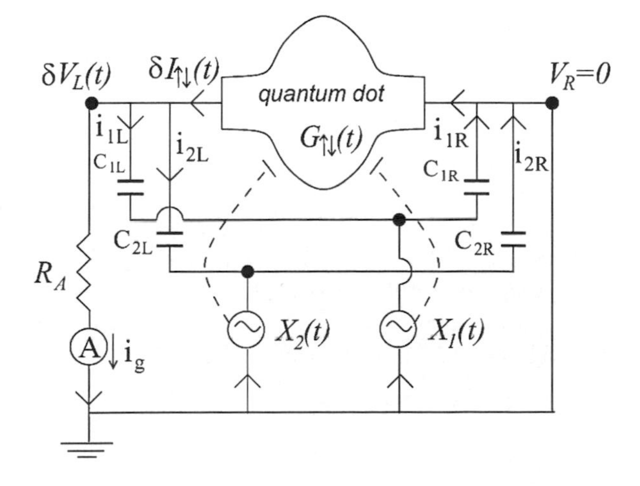

Fig. 7.6 Circuit model for the quantum pumping current measurement (current setup). The spurious capacitances between the electrode gates and the leads are denote by C_{1R}, C_{1L}, C_{2R}, and C_{2L}. R_A denotes the ammeter internal resistance

where C incorporates the characteristics of the dot and model circuit. Notice that this expression is very similar to (7.30). Indeed, using (7.34) instead of (7.30) to evaluate the polarization coefficient r_{pol} as a function of the magnetic field amplitude, we have found nearly identical curves to those shown in Fig. 7.4 [21]. Experimentally, it would be difficult to distinguish between pure quantum pumping and rectified spin currents on the basis of the magnetic field dependence of the correlations alone. Moreover, it turns out that even pure spin currents can be generated through rectification using the circuit model show in Fig 7.6. Thus, the most effective way to find out by how much the spin pumping current is contaminated by rectification remains the symmetry analysis described in Sect. 7.2.2.

7.4.2 Dephasing

Orbital decoherence will limit the amplitude of spin pumping currents much in the same way it does for charge pumping. Since quantum pumping of charge is based on having phase coherence within the dot, it disappears altogether when the electron dephasing length is smaller than the dot linear size. In this case, driving the quantum dot parametrically may still produce currents, but the mechanisms involved are either merely classical or simply related to rectification.

In fact, the effect of decoherence is even more drastic in the case of spin transport. That is because decoherence also washes out the spatial features of wave functions, causing all states to carry very similar pumping currents. This makes the Zeeman splitting less effective in uncorrelating "up" and "down" components of the current, decreasing the amplitude of the spin current. Notice that the same is true even if rectification is the main source of spin currents: as wave function "speckles" are smeared by dephasing, $G_\uparrow(t)$ and $G_\downarrow(t)$ become very similar, reducing the amplitude of $\delta I_s = \overline{\delta I_\uparrow(t) - \delta I_\downarrow(t)}$ as well (the bar here denotes average over time).

We have studied how decoherence impacts spin pumping in Ref. [21]. We followed the approach pioneered by Büttiker [34] where dephasing inside the dot is accounted for within the scattering matrix formalism by adding a third lead to the dot. The third lead carries no net current and has the sole purpose of randomizing the electron phase. The number of channels in the third lead, N_ϕ, is taken to be very large, but their coupling to the dot, p, is assumed very small. We can adjust the product pN_ϕ to produce the desired dephasing rate.

Our results show that for dots with a large number of propagating channels, setting pN_ϕ to yield realistic dephasing rates does not impact the amplitude of the spin current fluctuations as much as thermal fluctuations do. Thus, for the experimental conditions found in Ref. [26], dephasing plays a minor role in comparison to thermal smearing.

7.5 Summary and Future Directions

In this lecture we have argued that adiabatic pumping in open, lateral quantum dots can be used effectively to generate spin-polarized currents. The method is based on the phase coherence and the sensitivity of wave functions and energy levels inside the dot to changes in the shape of the confining potential. Realistic estimates based on random-matrix theory were presented and critical shortcomes such as spurious rectification and decoherence were addressed. Recently, the proposal passed its first experimental test. We hope that more groups will become interested in this subject.

During our presentation, we also left out some aspects which deserve further investigation. First, it is clear that the spin currents generated by this method can only be polarized parallel to the dot planar structure. In order to achieve polarizations perpendicular to the plane or along any arbitrary direction one could explore the spin–orbit coupling intrinsic to asymmetric heterostructures (Rashba effect) and to semiconductors with no crystalline inversion symmetry (such as GaAs). A proposal along this direction already exists [24], but so far has not been implemented experimentally.

We have also not discussed spin relaxation as well as dissipation. There is strong evidence that spin-polarized currents travel considerable distances in GaAs 2DEG [35]. While pure dc spin currents lead to zero net charge or discharging, it is erroneous to believe that they are accompanied by no Joule dissipation. The electric power, when averaged over a full cycle, is nonzero. It is important to emphasize that dissipation increases substantially in the non-adiabatic regime.

A third aspect is how quantum pumping is affected by electron–electron interactions. It is well-known that the Coulomb interaction does not play an important role in transport properties of open quantum dots, namely, when channels in the leads are fully propagating [36, 37]. The Coulomb interactions becomes important when the contacts are pinched so strongly that electrons have to tunnel to enter or leave the quantum dot [12]. When the number of electrons in the dot is constrained to be an integer, charging effects take over and transport occurs only one electron at a time. In this case, the system resembles a classical turnstile pump and quantum interference plays essentially no role.

Although our proposed pump operates far from this regime, it would be very interesting to understand how one can interpolate between the fully open and the closed, Coulomb blockade regime. The difficulty lies in finding a formalism where both charging, many-body effects, phase coherence, and time-dependent driving can be put together. A few recent attempts along this direction have already been taken [38, 39]. Perhaps the most fascinating issue to be studied is the interplay between Kondo correlations and external driving. Such investigations might allow one to devise more efficient and controllable mechanisms to generate spin currents with quantum dots [40–42].

Acknowledgments I am grateful to C. Chamon, C. Lewenkopf, C. Marcus, and M. Martínez-Mares for fruitful collaborations on this subject. The material presented here is based on published work we have co-authored. I also would like to thank P. Brouwer, B. Reulet, P. Sharma, and S. Watson for useful discussions.

References

1. Kastner, M.A.: Rev. Mod. Phys. **64**, 849 (1992)
2. Marcus, C.M., Rimberg, A.J., Westervelt, R.M., Hopkins, P.F., Gossard, A.C.: Phys. Rev. Lett. **69**, 506 (1992)
3. Chang, A.M., Baranger, H.U., Pfeiffer, L.N., West, K.W., Chang, T.Y. : Phys. Rev. Lett. **76**, 1695 (1996)
4. Kouwenhoven, L.P., et al.: In: Kouwenhoven, L.P., Schön, G., Sohn, L.L. (eds.) Nato ASI Conference Proceedings. Kluwer, Dordrecht (1997)
5. Goldhaber-Gordon, D., Shtrikman, H., Mahalu, D., Abusch-Magder, D. Meirav, U., Kastner, M.A.: Nature **391**, 156 (1998)
6. Cronenwett, S.M., Oosterkamp, T.H., Kouwenhoven, L.P.: Science **281**, 540 (1998)
7. Switkes, M., Marcus, C.M., Campman, K., Gossard, A.C.: Science **283**, 1905 (1999)
8. Thouless, D.J.: Phys. Rev. B **27**, 6083 (1983)
9. Altshuler, B.L., Glazman, L.I.: Science **283**, 1864 (1999)
10. Spivak, B., Zhou, F., Beal Monod, M.T.: Phys. Rev. B **51**, 13226 (1995)
11. Brouwer, P.W.: Phys. Rev. B **58**, R10135 (1998)
12. Aleiner, I.L., Andreev, A.V.: Phys. Rev. Lett. **81**, 1286 (1998)
13. Zhou, F., Spivak, B., Altshuler, B.L.: Phys. Rev. Lett. **82**, 608 (1999)
14. Wolf, S.A., et al.: Science **294**, 1488 (2001)
15. Zutic, I., Fabian, J., Das Sarma, S.: Rev. Mod. Phys. **76**, 323 (2004)
16. Büttiker, M., Prêtre, A., Thomas, H.: Z. Phys. B **94**, 196 (1994)
17. Avrom, J.E., Elgart, A., Graf, G.M., Sadum, L.: J. Stat. Phys. **116**, 425 (2004)
18. Brouwer, P.W.: Phys. Rev. B **63**, 121303(R) (2001)
19. Beenakker, C.W.J.: Rev. Mod. Phys. **69**, 731 (1997)
20. Cremers, J.N.H.J., Brouwer, P.W.: Phys. Rev. B **65**, 115333 (2002)
21. Martínez-Mares M., Lewenkopf C.H., Mucciolo E.R.: Phys. Rev. B **69**, 085301 (2004)
22. Aleiner, I.L., Altshuler, B.L., Kamenev, A.: Phys. Rev. B **62**, 10373 (2000)
23. Governale, M., Taddei, F., Fazio, R.: Phys. Rev. B **68**, 155324 (2003)
24. Sharma, P., Brouwer, B.W.: Phys. Rev. Lett. **91**, 166801 (2004)
25. Mucciolo, E.R., Chamon, C., Marcus, C.M.: Phys. Rev. Lett. **89**, 146802 (2002)
26. Watson, S.K., Potok, R.M., Marcus, C.M., Umansky, V.: Phys. Rev. Lett. **91**, 258301 (2003)
27. Sharma, P., Chamon, C.: Phys. Rev. Lett. **87**, 096401 (2001)
28. Citro, R., Andrei, N., Niu, Q.: Phys. Rev. B **68** 165312 (2003)
29. Simons, B.D., Altshuler, B.L.: In *Mesoscopic Quantum Physics*, Eds. E. Akkermans, G. Montambaux, J.-L. Pichard, J. Zinn-Justin (Elsevier, 1994)
30. Vavilov, M.G., Ambegaokar, V., Aleiner, I.L.: Phys. Rev. B **63**, 195313 (2001)
31. Simons, B.D., Altshuler, B.L.: Phys. Rev. Lett. **70**, 4063 (1993)
32. Folk, J.A., Potok, R.M., Marcus, C.M., Umansky, V.: Science **299**, 679 (2003)
33. van Wees, B.J., et al.: Phys. Rev. Lett. **60**, 848 (1988)
34. Büttiker, M.: Phys. Rev. B **33**, 3020 (1986)
35. Kikkawa, J.M., Awschalom, D.D.: Phys. Rev. Lett. **80**, 4313 (1998)
36. Brouwer, P.W., Aleiner, I.L.: Phys. Rev. Lett. **82**, 390 (1999)
37. Aleiner, I.L., Brouwer, P.W., Glazman, L.I.: Phys. Rep. **358**, 309 (2002)
38. Splettstoesser, J., Governale, M., König, J., Fazio, R.: Phys. Rev. Lett. **95**, 246803 (2005)
39. Sela, E., Oreg, Y.: Phys. Rev. B **71(7)**, 075323 (2005)
40. Feinberg, D., Simon, P.: Appl. Phys. Lett. **85**, 4247 (2004)
41. Blaauboer, M., Fricot, C.M.L.: Phys. Rev. B **71**, 041303 (2005)
42. Cota, E., Aguado, R., Platero, G.: Phys. Rev. Lett. **94**, 107202 (2005)

Chapter 8
Spin Relaxation: From 2D to 1D

Alexander W. Holleitner

8.1 Introduction

Semiconductor spintronics seeks to gain extra functionality compared to conventional electronics by exploiting the carrier spin degree of freedom [1, 2] (Fig. 8.1). For a potential processing scheme, which combines quantum mechanical and classical information, it is of particular interest to manipulate and to control carrier spin dynamics in non-magnetic materials by utilizing the spin–orbit interaction [3–6]. Datta and Das [7] proposed the concept of a spin-polarized field-effect transistor in narrow-band-gap semiconductors such as InGaAs. On the one hand, the structure inversion asymmetry in these materials allows controlling the electron spin precession by an electric field [8–14], and it appears when the layers of a semiconductor heterostructure are not symmetrically arranged with respect to the conducting layer. On the other hand, the range of operation of the spin transistor is limited by the spin relaxation mechanism induced by the spin–orbit interaction [15–17]. Here, we describe the possibility of an effective suppression of the spin relaxation in the two-dimensional conducting channel of a spin transistor [18–21]. As a precursor of the one-dimensional limit, long spin relaxation times are expected when the channel width w is smaller than the electronic mean free path l_e ($w < l_e$). As a result, spin relaxation mechanisms, which are intrinsic to the two-dimensional electron systems (2DES), can be suppressed very efficiently [22].

A. W. Holleitner (✉)
Walter Schottky Institut and Physik-Department, Technische Universität München,
Am Coulombwall 3, 85748 Garching, Germany
e-mail: holleitner@wsi.tum.de

Vojta et al. (eds.), *CFN Lectures on Functional Nanostructures – Volume 2*,
Lecture Notes in Physics 820, DOI: 10.1007/978-3-642-14376-2_8,

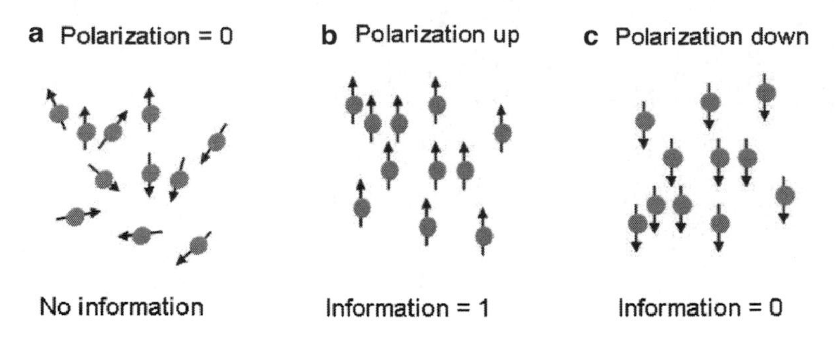

Fig. 8.1 In semiconductor spintronics, information is encoded via the spin polarization of the conduction electrons. **a** If the distribution of the electron spin is in equilibrium, no information is encoded. A non-equilibrium spin polarization can carry the information **b** one or **c** zero

8.2 The Spin–Orbit Interaction in Zinc-Blende Structures

In inversion asymmetric heterostructures, the spin–orbit interaction is determined by the interplay between the bulk inversion asymmetry (BIA) and the structural inversion asymmetry (SIA) and the interface asymmetry (IA) [23–26]. For zinc-blende crystals, the corresponding effective mass Hamiltonan can be written as [27]

$$H = \tilde{\eta} \cdot [\sigma_x k_x (k_y^2 - k_z^2) + cycl.perm.], \tag{8.1}$$

where $\sigma_{x,y,z}$ are the Pauli matrices, and $\tilde{\eta}$ is a constant which reflects the strength of the spin–orbit interaction in the conduction band. In the case of a quasi two-dimensional quantum well, the wave vector k_z of allowed states along the growth direction is typically much larger than the two in-plane components k_x and k_y. To describe the spin–orbit interactions in a semiconductor quantum well correctly, energy terms with higher orders in k need to be considered [28]. However, the energy terms which are linear in k are sufficient for the discussion below. Usually two Hamiltonians linear in k are assumed to describe the spin–orbit interactions in a quantum well. The corresponding Hamiltonian, which refers to the BIA in a quantum well, was first described by Dresselhaus [29]

$$H_{\mathrm{BIA}} = \check{\eta} \cdot [\sigma_x k_x + \sigma_y k_y]. \tag{8.2}$$

In principle, the Dresselhaus Hamiltonian is derived from (8.1), by considering only terms with k_z^2 [19]. In asymmetric quantum wells and deformed bulk systems, the structure inversion asymmetry gives rise to the Bychkov–Rashba spin orbit coupling [30–32]

$$H_{\mathrm{SIA}} = \hat{\eta} \cdot [\sigma_x k_y - \sigma_y k_x] = \hat{\eta}\boldsymbol{\sigma} \cdot [\mathbf{k} \times \mathbf{z}]. \tag{8.3}$$

If an angular frequency $\boldsymbol{\Omega}$ is defined as $\boldsymbol{\Omega}(\mathbf{k}) \equiv \alpha \cdot [\mathbf{v}(\mathbf{k}) \times \mathbf{z}]$ with $\mathbf{v}(\mathbf{k}) = \hbar\,\mathbf{k}/m^*$ the electron velocity, this Hamiltonian can be rewritten as

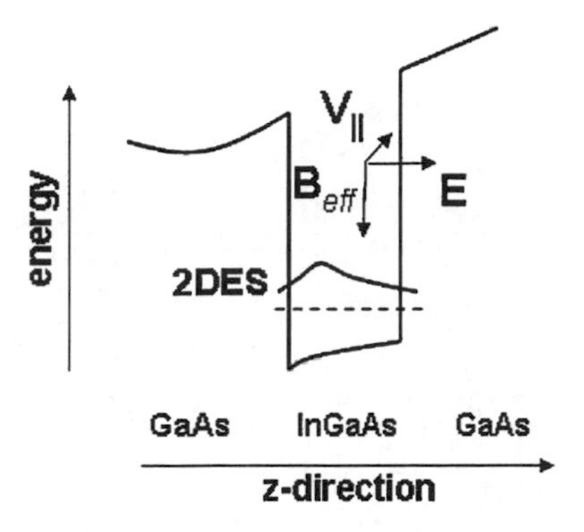

Fig. 8.2 Schematics of the conduction band of a heterostructure containing an asymmetric quantum well. In the example, a two-dimensional electron system (2DES) is formed in an InGaAs layer in between two GaAs layers. If an electric field **E** is applied along the growth direction of the heterostructure, the resulting spin–orbit interaction induces an effective magnetic field $\mathbf{B}_{\mathrm{eff}}(\mathbf{k})$ which is orientated in the plane of the quantum well and perpendicular to the velocity $\mathbf{v}_{||}(\mathbf{k})$ of the electron

$$H_{\mathrm{SIA}} = (\hbar/2)\boldsymbol{\sigma} \cdot \boldsymbol{\Omega}(\mathbf{k}). \tag{8.4}$$

In this way, the energy splitting for electrons with opposite spin directions can be regarded as an effective magnetic field $\mathbf{B}_{\mathrm{eff}}$ ($\mathbf{k}$), which depends on the wave vector of the electron. During a ballistic flight, the spin of an electron will precess with the angular frequency $\boldsymbol{\Omega}(\mathbf{k})$ while $1/\sigma$ defines the length, at which the spin of the electron rotates by an angle of π. Figure 8.2 depicts the situation in which an electric field is applied across an asymmetric quantum well. The resulting SIA of the quantum well gives rise to the effective magnetic field which is oriented in-plane. For asymmetric GaAs/Al$_x$Ga$_{1-x}$As quantum wells with an electron concentration of 10^{12} cm^{-2}, the spin splitting in the conduction band is measured to be of the order of 0.2–0.3 meV at the Fermi energy, corresponding to $\alpha = 6$–9×10^4 cm^{-1} [19, 33]. The orientation of the effective magnetic fields due to the BIA and the SIA could recently be characterized by optoelectronic measurements, in accordance with (8.2) and (8.3) [12].

8.3 The Datta–Das Spin Transistor

Based upon the voltage tunable Rashba Hamiltonian of (8.3), Datta and Das proposed the concept of a spin-polarized field effect transistor in narrow-band-gap semiconductors such as InGaAs [7, 34]. In such a spin transistor, the source and drain contacts are made of ferromagnetic materials (Fig. 8.3). As demonstrated by recent experiments [35–37], such contacts inject and collect preferentially spin polarized electrons; they act as spin polarizers and detectors. The injected spin-polarized carriers precess in transit through the 2DES due to the Rashba effect.

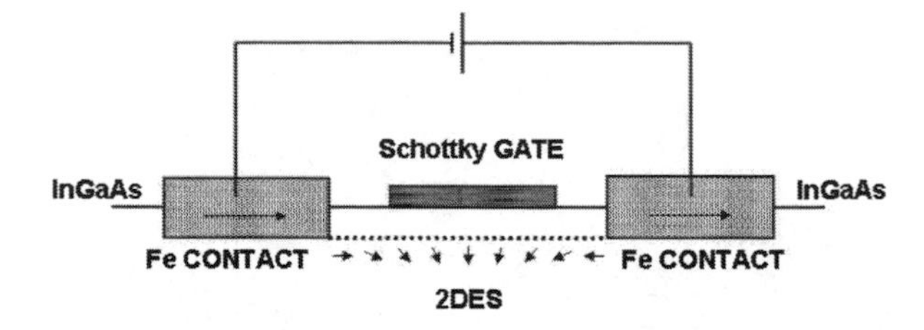

Fig. 8.3 Spin field-effect transistor, redrawn from ref. [7]. Spin polarization is injected into a 2DES via a ferromagnetic contact. The spin precession is controlled by the application of a voltage to the Schottky gate, while the final orientation of the electron spin is read-out by the second ferromagnetic contact. *Black arrows* denote the orientation of the electron spin

The transverse electric field can be tuned by applying a voltage to the Schottky gate on top of the heterostructure. If the magnetization in the drain and source contacts are parallel to each other, and the carriers perform an integral number of precessions, then the device conductance is high. In the case of opposite orientation for the electron spin and the drain magnetization, the device conductance is minimized. In addition to the gate voltage, the transconductance can also be controlled via an external magnetic field. The spin transistor initiated many studies on spin-polarized transport and spin-injection phenomena in semiconductor/ferromagnetic junctions [8–12, 14–17].

8.4 Spin Relaxation Mechanisms in Semiconductors

The spin transistor relies upon controlling the precession of the electron spin in the conducting channel due to the influence of the Rashba term. If an electron is being scattered in multiple events, the orientation of $\mathbf{\Omega}(\mathbf{k})$ changes randomly. Since the elementary rotations do not commute in the plane of a quantum well (nor in the case of a three-dimensional system), the final spin orientation varies for an electron being scattered along two different trajectories from one point in space to another. Considering an ensemble of trajectories, the total spin polarization is randomized after a certain number of scattering events. The corresponding spin relaxation mechanism is named after D'yakonov and Perel' [38]. In the motional narrowing regime, i.e., when the spin rotation angle is small during the ballistic flight of an electron $(\tau_p \Omega < 1)$, the spin relaxation time τ_S is inversely proportional to the momentum scattering time τ_P, i.e., $\tau_S{}^{-}1 \propto \tau p\, \Omega^2\,(k)$ [39]. In Fig. 8.4a, the spin precession is sketched for an electron moving along a certain trajectory. Every time the electron is scattered, the orientation of $\mathbf{\Omega}$ changes. Another source of spin relaxation in spin-orbit materials is given by the fact that spin relaxation due to momentum relaxation is possible directly through the spin–orbit coupling [40]. Generally, the Bloch states are not eigenstates of the Pauli matrixes. Thus, a lattice-induced spin–orbit interaction, e.g., influenced by phonons, can directly

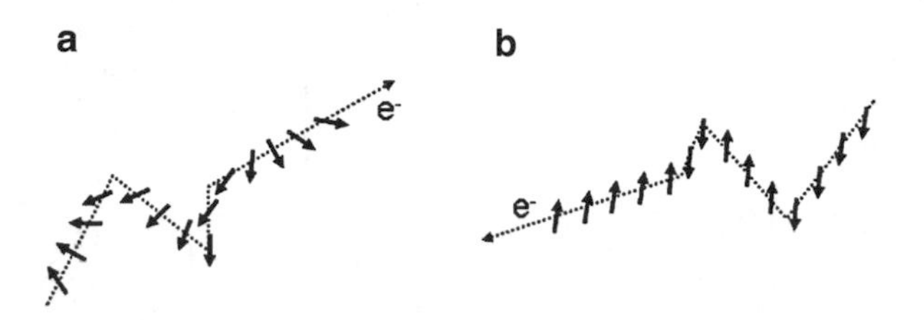

Fig. 8.4 Schematics of electron trajectories (*dotted lines*) in the case of **a** the D'yakonov–Perel' mechanism and **b** the Elliot–Yafet mechanism. The orientation of the spin eigenfunction is presented with *black arrows*

couple the (Pauli) spin up states to the spin down states [41]. Similarly, scattering at impurities and boundaries of the electron system can lead to instantaneous spin flip events (Fig. 8.4b) [42]. A further spin-relative relaxation mechanism was suggested by Bir et al. [43]. A simultaneous spin flip of electron and hole spin, which is intermediated by the electron hole exchange interaction, can induce an effective spin relaxation of the electron spin polarization. Last but not least, the hyperfine interaction can also cause a spin flip between electrons in the conduction band and the nuclear magnetic moments of the crystal. The latter is of particular relevance for heterostructures with a nuclear magnetic moment, such as GaAs (see ref. [17] for a detailed review on spin relaxation processes in semiconductors).

8.5 Transition from 2D to 1D

In two and three dimensions the elementary rotations do not commute. As a result, scattering processes tend to randomize the spin polarization (see Sect. 8.4). Bournel et al. [18] were the first to consider a 2DES with a finite width as the conducting channel of a spin transistor. They studied an asymmetric quantum well made of $In_{0.53}Ga_{0.47}As$; a system well known for its spin–orbit interactions due to SIA [8]. As a precursor of the one-dimensional limit, the results suggest long spin relaxation times, when the channel width w is comparable to the electronic mean free path l_e ($w \leq l_e$). Bournel et al. used a Monte-Carlo transport model which is based upon the collective motion of individual particles. In between instantaneous scattering events, e.g., with phonons and alloy impurities, the ballistic flight of an electron is only determined by the influence of the Rashba term. For their simulation, the authors assumed specular reflection at the boundaries of the conducting channel. At the same time, the two spin populations, injected by the source contact, are supposed to be non-intermixing. The latter condition means that other spin relaxation mechanisms than the D'yakonov–Perel' mechanism, such as the one by the Elliott–Yafet, are neglected. As a result, their simulations suggest almost negligible spin relaxation in the conducting channel of an $In_{0.53}Ga_{0.47}As$ spin

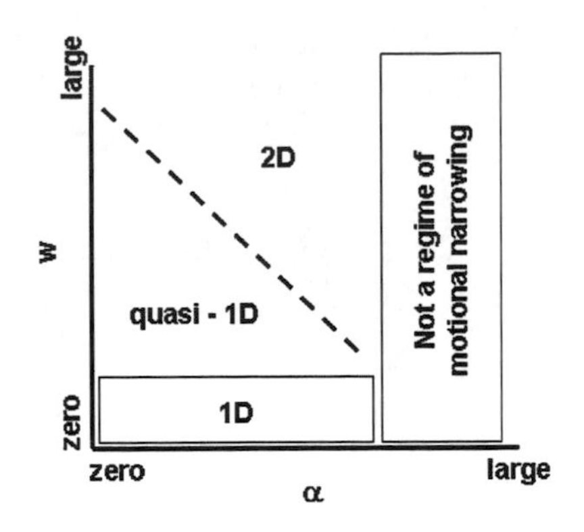

Fig. 8.5 Phase diagram of the transition 2D to 1D for a spin transistor in dependence of the finite width w of a 2DES and the spin–orbit constant α (after ref. [19].) See text for details

transistor for widths of the order of 100 nm and electric fields smaller than 200 kV/cm [18].

In a similar Monte-Carlo transport study, Kiselev and Kim identified different regimes of the spin relaxation in a 2DES with a finite width (see Fig. 8.5) [19]. The authors assume in their work that the D'yakonov–Perel' mechanism is the dominant spin relaxation mechanism in a 2DES, in which the spin–orbit interaction is dominated by the SIA term. For a very large spin splitting constant, i.e., $l_e\,\alpha \geq 1$, the assumptions for motional narrowing are not fulfilled anymore (see Sect. 8.2 for the definition of α). Here, scattering events randomize the spin information, since the precession angle of the electron spin is large during the ballistic flight of the electron ($\tau_P\Omega > 1$). In turn, spin relaxation time is proportional to the momentum scattering time $\tau_S \sim \tau_P$. In the motional narrowing regime, where $l_e\alpha < 1$, the spin relaxation time is given by $\tau_S^{2D} \sim \tau_P(l_e\alpha)^{-2}$ (upper range in Fig. 8.5 which is marked as '2D'). The authors claim that if the width w of the conducting channel is reduced, a transition occurs at $w\alpha \sim 1$ (marked as a dashed line in Fig. 8.5). Kiselev and Kim find that this transition can occur at widths which are ten times larger than the mean free path. For $w\alpha < 1$, the width of the channel acts as an effective mean free path of the system. In this quasi one-dimensional regime, the spin relaxation is suppressed very efficiently as $\tau_S \sim \tau_S^{2D}(w\alpha)^{-2}$.

Another approach to the problem was given by Mal'shukov and Chao [20], by solving the spin diffusion equation for electrons in the conducting channel. Their solutions showed that there are certain spin diffusion modes with slow spin relaxation rates if the conducting channel of the spin transistor has a finite width. In their model, they assume that the electron motion perpendicular to the channel can be described semi-classically. On the one hand, the width w needs to be much shorter than the spin precession length $1/\alpha$, and on the other hand, it needs to be larger than the Fermi wavelength. Similar to ref. [18, 19], Mal'shukov and Chao neglect the BIA, because it is assumed that in narrow gap semiconductors, the spin dynamics are dominated by the SIA. As a main result, the spin relaxation of the waveguide modes is slowed down as $\tau_s \sim (1/w\alpha)^2$, in accordance with the simulation of ref. [19]. In addition, the authors point out that the BIA term, being cubic in k, gives rise to an additional spin relaxation which is independent of the channel width w. Thus, the slowing-down and finally the suppression of the spin relaxation is limited by this cubic BIA term. In a recent publication, Winkler calculated the magnitude and the orientation of B_{eff} for an infinite two-dimensional InGaAs quantum well by considering higher order terms of the electron wave-vector [28]. Combining both results for semi-classical wires, one expects an anisotropy of the spin relaxation times for electrons moving into different crystallographic directions.

Pareek and Bruno [21] used a recursive Green's function method, which takes into account the quantum effects at the single particle level. In the case of motional narrowing, the authors again see an enhancement of the spin polarization, if the electrons are confined to a channel width of the order of the mean free path. However, they can not corroborate the result predicted by the real-space Monte Carlo simulations, that this enhancement should be detectable even for widths ten times the mean free path. In the quasi-ballistic regime, i.e., $l_e\alpha >> 1$, Pareek and Bruno find that the spin diffusion length can be even longer than the spin precession length. At the same time, the spin relaxation rates depend on the initial orientation of the electron spin. As a result, the largest current modulation can be obtained, if the electron spin is injected perpendicular to the plane of the quantum well of the spin transistor.

Generally speaking, all theory work, which covers the effect of the transition from 2D to 1D on the spin relaxation, predicts increasing spin relaxation times as a result of a dimensionally constrained D'yakonov–Perel' mechanism due to the SIA term. However, as Mal'shukov et al. pointed out, other mechanisms can limit the eventual suppression of the spin relaxation, e.g., the BIA induces a component of the D'yakonov–Perel' spin relaxation which is independent of the channel width [20]. We would like to note that in a first experimental study, an effective slowing of the D'yakonov–Perel' spin relaxation mechanism has been observed in conducting channels of n-InGaAs quantum wires [22, 44]. The results are consistent with a dimensionally-constrained D'yakonov–Perel' mechanism. For the narrowest wires with a width of only a few hundred nanometers, the authors find that an interplay between the spin diffusion length and the wire width determines the spin dynamics, and that channels along the crystallographic directions [100] and [010] show longer spin relaxation times than channels along [110] and [−110]. The anisotropy can be interpreted such that the cubic spin–orbit coupling terms due to BIA start to dominate the spin relaxation in the narrowest channels [24–26, 28]. We would like to further point out that a similar dimensional crossover has been observed by means of a weak anti-localization analysis of magnetotransport studies on InAs channels [45–49], and by means of Shubnikov-de Haas oscillations in narrow InAs-based heterostructures [50]. Based upon these experimental results, recent theory work substantiates the interpretation that the dimensional crossover can be understood in terms of an interplay between the channel width, the spin precession length, inter-subband scattering, and the effect of spin scattering at the boundaries of the channels [51–57].

8.6 Summary

In semiconductors without inversion symmetry, the precession of the electron spin can be influenced via the spin–orbit interaction. At the same time, spin relaxation mechanisms based on the spin–orbit coupling limit the range of control on the spin dynamics. Particularly in 3D and 2D electron systems, momentum scattering gives rise to a spin relaxation mechanism which was proposed by D'yakonov and Perel'. We describe recent theoretical results, which suggest that the D'yakonov–Perel' mechanism can be suppressed by decreasing the width of a two-dimensional electron system down to the electronic mean free path. In such quasi one-dimensional systems long spin relaxation times are expected.

Acknowledgments We would like to thank D.D. Awschalom, H. Knotz, F. Meier, V. Sih, R.C. Myers, A.C. Gossard, V. Khrapey, J.P. Kotthaus for discussions and support. We gratefully acknowledge financial support from the Center for NanoScience (CeNS) in Munich, Germany, the ONR, DARPA, and the California NanoSystems Institute (CNSI) at the University of California, Santa Barbara, USA, the DFG Project No. HO-3324/4, and the German Excellence Initiative via "Nanosystems Initiative Munich (NIM)" and "LMUexcellent".

References

1. Prinz, G.A.: Science **282**, 1660 (1998)
2. Wolf, S.A., Awschalom, D.D., Buhrman, R.A., Daughton, J.M., von Molnar, S., Roukes, M.L., Chtchelkanova, A.Y., Treger, D.M.: Science **294**, 1488 (2001)
3. Lommer, G., Malcher, F., Rössler, U.: Phys. Rev. Lett. **60**, 728 (1988)
4. Luo, J., Munekata, H., Fang, F.F., Stiles, P.J.: Phys. Rev. B **38**, 10142 (1988)
5. Luo, J., Munekata, H., Fang, F.F., Stiles, P.J.: Phys. Rev. B **41**, 7685 (1990)
6. Kato, Y., Myers, R.C., Myers, A.C., Awschalom, D.D.: Nature **427**, 50 (2003)
7. Datta, S., Das, B.: Appl. Phys. Lett. **56**, 665 (1989)
8. Nitta, J., Akazaki, T., Takayanagi, H., Enoki, T.: Phys. Rev. Lett. **78**, 1335 (1997)
9. Grundler, D.: Phys. Rev. Lett. **84**, 6074 (2000)
10. Koga, T., Nitta, J., Akazaki, T., Takayanagi, H.: Phys. Rev. Lett. **89**, 046801 (2002)
11. Miller, J.B., Zumbühl, D.M., Marcus, C.M., Lyanda-Geller, Y.B., Goldhaber-Gordon, D., Campman, K., Gossard, A.C.: Phys. Rev. Lett. **90**, 076807 (2003)
12. Ganichev, S.D., Bel'kov, V.V., Golub, L.E., Ivchenko, E.L., Schneider, P., Giglberger, S., Eroms, J., de Boeck, J., Borghs, G., Wegscheider, W., Weiss, D., Prettl, W.: Phys. Rev. Lett. **92**, 256601 (2004)
13. Schäpers, Th., Knobbe, J., Guzenko, V.A.: Phys. Rev. B **69**, 235323 (2004)
14. Hansen, A.E., Björk, M.T., Fasth, C., Thelander, C., Samuelson, L.: Phys. Rev. B **71**, 205328 (2005)
15. Averkiev, N.S., Golub, L.E.: Phys. Rev. B **60**, 15582 (1999)
16. Schliemann, J., Egues, J.C., Loss, D.: Phys. Rev. Lett. **90**, 146801 (2003)
17. Zutic, I., Fabian, J., Das Sarma, S.: Rev. Mod. Phys. **76**, 323 (2004)
18. Bournel, A., Dollfus, P., Bruno, P., Hesto, P.: Eur. Phys. J. AP **4**, 1 (1998)
19. Kiselev, A.A., Kim, K.W.: Phys. Rev. B **61**, 13115 (2000)
20. Mal'shukov, A.G., Chao, K.A.: Phys. Rev. B **61**, 2413 (2000)
21. Pareek, T.P., Bruno, P.: Phys. Rev. B **65**, 241305 (2002)

22. Holleitner, A.W., Sih, V., Myers, R.C., Gossard, A.C., Awschalom, D.D.: Phys. Rev. Lett. **97**, 036805 (2006)
23. Heida, J.P., van Wees, B.J., Kuipers, J.J., Klapwijk, T.M., Borghs, G.: Phys. Rev. B **57**, 11911 (1998)
24. Ting, D.Z.-Y., Cartoixa, X.: Phys. Rev. B **68**, 235320 (2003)
25. Lusakowski, A., Wrobel, J., Dietl, T.: Phys. Rev. B **68**, 081201 (2003)
26. Liu, M.-H., Chang, C.-R., Chen, S.-H.: Phys. Rev. B **71**, 153305 (2005)
27. D'yakonov, M.I., Perel, V.I.: Sov. Phys. JETP **33**, 1053 (1971)
28. Winkler, R.: Phys. Rev. B **69**, 045317 (2004)
29. Dresselhaus, G.: Phys. Rev. **100**, 580 (1955)
30. Rashba, E.I.: Sov. Phys. Solid State **2**, 1109 (1960)
31. Bichkov, Yu.A., Rashba, E.I.: J. Phys. C **17**, 6039 (1984)
32. Bichkov, Yu.A., Rashba, E.I.: JETP Lett. **39**, 78 (1984)
33. Wissinger, L., Rössler, U., Winkler, R., Jusserand, B., Richards, D.: Phys. Rev. **58**, 15375 (1998)
34. Knap, W., Skierbiszewski, C., Zduniak, A., Litwin-Staszewska, E., Bertho, D., Kobbi, F., Robert, J.L., Pikus, G.E., Pikus, F.G., Iordanskii, S.V., Mosser, V., Zekentes, K., Lyanda-Geller, Y.B.: Phys. Rev. B **53**, 3912 (1996)
35. Hammar, P.R., Bennett, B.R., Yang, M.J., Johnson, M.: Phys. Rev. Lett. **83**, 203 (1999)
36. Stephens, J., Berezovsky, J., McGuire, J.P., Sham, L.J., Gossard, A.C., Awschalom, D.D.: Phys. Rev. Lett. **93**, 097602 (2004)
37. Crooker, S.A., Furis, M., Lou, X., Adelmann, C., Smith, D.L., Palmstrøm, C.J., Crowell, P.A.: Science **309**, 2191 (2005)
38. Dyakonov, M.I., Perel', V.I.: Sov. Phys. Solid State **13**, 3023 (1971)
39. Lau, W., Olesberg, J.T., Flatte, M.E.: Phys. Rev. B **64**, 161301 (2001)
40. Elliott, P.G.: Phys. Rev. **96**, 266 (1954)
41. Yafet, Y.: In: Seitz, F., Turnbull, D. (eds.) Solid State Physics, vol 13, (Academic, New York 1963)
42. Fabian, J., Das Sarma, S.: J. Vac. Sci. Technol. B **17**, 1708 (1999)
43. Bir, G.L., Aronov, A.G., Pikus, G.E.: Zh. Eksp. Teor. Fiz. **69**, 1382 (1975)
44. Holleitner A.W., Sih V., Myers R.C., Gossard A.C., Awschalom D.D.: New J. Phys. **9**, 342 (2007)
45. Wirthmann, W., Gui, Y.S., Zehnder, C., Heitmann, D., Hu, C.M., Kettemann, S.: Phys. E **34**, 493 (2006)
46. Schäpers, Th., Guzenko, V.A., Pala, M.G., Zülicke, U., Governale, M., Knobbe, J., Hardtdegen, H.: Phys. Rev. B **74**, 081301(R) (2006)
47. Schäpers, Th., Guzenko, V.A., Bringer, A., Akabori, M., Hagedorn, M., Hardtdegen, H.: Semicond. Sci. Technol. **24**, 064001 (2009)
48. Kunihashi, Y., Kohda, M., Nitta, J.: Phys. Rev. Lett. **102**, 226601 (2009)
49. Nitta, J., Bergsten, T., Kunihashi, Y., Kohda, M.: J. Appl. Phys. **105**, 122402 (2009)
50. Kwon, J.H., Koo, H.C., Chang, J., Han, S.-H., Eom, J.: Appl. Phys. Lett. **90**, 112505 (2007)
51. Schwab, P., Dzierzawa, M., Gorini, C., Raimondi, R.: Phys. Rev. B **74**, 155316 (2006)
52. Kettemann, S.: Phys. Rev. Lett. **98**, 176808 (2007)
53. Dragomirova, R.L., Nikolic, B.K.: Phys. Rev. B **75**, 085328 (2007)
54. Kaneko, T., Koshino, M., Ando, T.: Phys. Rev. B **78**, 245303 (2008)
55. Scheid, M., Kohda, M., Kunihashi, Y., Richter, K., Nitta, J.: Phys. Rev. Lett. **101**, 266401 (2008)
56. Chang, C.-H., Tsai, J., Lo, H.-F., Mal'shukov, A.G.: Phys. Rev. B **79**, 125310 (2009)
57. Lü, C., Schneider, H.C., Wu, M.W.: J. Appl. Phys. **106**, 073703 (2009)

Chapter 9
Electronic Transport Properties of Superconductor–Ferromagnet Hybrid Structures

Detlef Beckmann

9.1 Introduction

Superconductivity and ferromagnetism are in most cases antagonistic long-range orders. In conventional singlet superconductors, the electrons have a pairwise antiparallel spin alignment, whereas ferromagnetism prefers parallel spin alignment. This prevents the coexistence of ferromagnetism and superconductivity in bulk materials, except for some exotic materials with unconventional superconducting pairing symmetry. In hybrid structures, however, the competition of superconductivity and ferromagnetism leads to a variety of interesting phenomena, some of which have potential applications in spintronics or quantum information processing devices. The typical intrinsic length scales of superconductivity and ferromagnetism are in the nanometer regime, and the advent of nanofabrication techniques has boosted the experimental advance in the field.

Here, we will focus on electronic transport properties of superconductor–ferromagnet (S/F) hybrid structures. Equilibrium properties will be presented only where they elucidate the main subject. The article is organized as follows: First, current transport through a single interface between a superconductor and a ferromagnet will be treated. Then, structures with two S/F interfaces will be discussed, namely S/F/S Josephson junctions and F/S/F spin-valve structures. For each section, an introduction to the corresponding superconductor/normal metal structures (S/N) will be given, and then the changes due to ferromagnetism are discussed. The aim is to give a self-contained picture of the relevant physical mechanisms, which will be presented along with recent experimental highlights.

D. Beckmann (✉)
Institut für Nanotechnologie, Karlsruhe Institut für Technologie (KIT), Postfach 3640, 76021 Karlsruhe, Germany
e-mail: detlef.beckmann@kit.edu

Vojta et al. (eds.), *CFN Lectures on Functional Nanostructures – Volume 2*, Lecture Notes in Physics 820, DOI: 10.1007/978-3-642-14376-2_9,

The theoretical background will be kept at an elementary level, and the reader is referred to specialized reviews for further details.

9.1.1 Ferromagnetism

The electronic band structure of paramagnetic metals is spin-degenerate, and in its simplest form can be modeled by a free-electron parabola. In an applied magnetic field, spin degeneracy is lifted, and the sub-bands for spin-up and spin-down are rigidly shifted by the Zeeman energy, leading to Pauli paramagnetism. The itinerant ferromagnetism of transition metals can be described by the Stoner model [1]. It is based on the above picture, with the addition of the exchange interaction, yielding an intrinsic exchange field h_{ex} which acts like the Zeeman energy (Fig. 9.1). The typical energy scale is $h_{ex} \approx 1$ eV, i.e., a sizeable fraction of the Fermi energy ϵ_F. A very interesting special case are half-metallic ferromagnets, which have only one spin species delocalized at the Fermi energy, and in terms of our simple model are characterized by $h_{ex} \geq \epsilon_F$. Two consequences of the spin-split dispersion relation relevant here are that the density of states at the Fermi energy and the Fermi wave vector are spin-dependent. For a practical quantitative analysis of experiments, it is usual to define a degree of spin polarization in the form $P_X = (X^\uparrow - X^\downarrow)$ $(X^\uparrow + X^\downarrow)$, where X is some spin dependent quantity related to ferromagnetism, like the density of states at the Fermi energy or conductivity.

9.1.2 Superconductivity

Superconductivity is characterized by the formation of electron pairs, the Cooper pairs, which are bosonic and condense into a macroscopic quantum state, thereby eliminating single-particle scattering and yielding a vanishing electric resistance

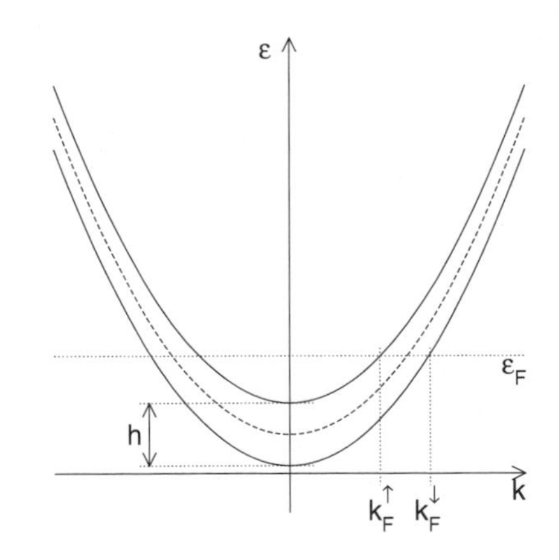

Fig. 9.1 Simplified model of the dispersion relation in a ferromagnet. The band structure is split into spin-up and spin-down bands, shifted rigidly by the exchange field h_{ex}

[2]. For conventional superconductors, the binding energy of the Cooper pairs is provided by electron–phonon interaction, and they are formed of electrons of opposite spin and momentum, described by the theory of Bardeen, Cooper and Schrieffer (BCS) [3]. The single-particle excitation spectrum exhibits an energy gap, $\Delta = 1.76\ k_B T_c$ for weak electron–phonon coupling at $T = 0$, where k_B is the Boltzmann constant and T_c is the critical temperature. There is a finite length scale, ξ_S, the coherence length, over which superconducting order, i.e., the spatial density of Cooper pairs can vary. Typical energy and length scales are $\Delta = 0.1$–1 meV and $\xi_S = 100$–$1{,}000$ nm, respectively.

9.2 Single S/F Interfaces

Charge transport through the interface between a normal metal and a superconductor is mediated by two elementary processes: quasiparticle tunneling and Andreev reflection.

9.2.1 *Tunneling*

Tunneling of individual quasiparticles between a normal metal and a superconductor can be described in terms of the so-called semiconductor model of superconductivity, shown in Fig. 9.2. On the right hand side, a scheme of the quasiparticle density of states in the superconductor is shown. There are no states inside the energy gap from $-\Delta$ to Δ. Below this interval, all states are filled, whereas at positive energy, all states are empty. The density of states exhibits a

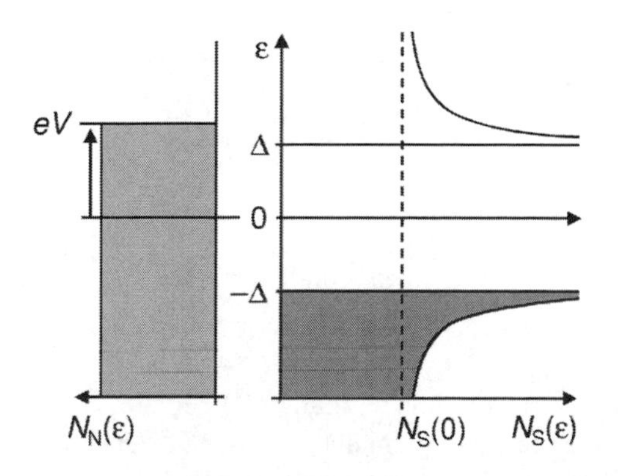

Fig. 9.2 Schematic view of quasiparticle tunneling in the semiconductor model of superconductivity. The *left hand side* shows the density of states $N_N(\epsilon)$ of the normal metal, shifted by the applied voltage. The *right hand side* is the superconducting quasiparticle density of states $N_S(\epsilon)$, with an energy gap Δ and sharp peaks near the gap edges. $\epsilon = 0$ refers to the chemical potential of the superconductor

sharp peak near the gap edges, and falls off to the (constant) normal state density of states $N_S(0)$ at high energies. On the left hand side, the density of states in the normal metal, offset by the applied voltage eV, is shown. Tunneling occurs horizontally from occupied states of the normal metal to unoccupied states in the superconductor (and vice-versa). The occupation probabilities are described by Fermi functions. The total current is found by integrating over energy, and is given by

$$I_{NS} = \alpha N_N(0) N_S(0) |\tau|^2 \int_{-\infty}^{\infty} n_{BCS}(\epsilon)[f(\epsilon - eV) - f(\epsilon)] d\epsilon, \tag{9.1}$$

where α is a proportionality constant, τ the tunneling-matrix element between N and S, e the electron charge, $n_{BCS}(\epsilon) = |\epsilon|/\sqrt{\epsilon^2 - \Delta^2}$ is the normalized BCS density of states, and $f(\epsilon)$ is the Fermi function. The factor 2 is included because we take the density of states per spin direction for later purposes. The information obtained from tunneling experiments can be seen most clearly in the differential conductance, given by

$$\frac{dI_{NS}}{dV} = G_{NN} \int_{-\infty}^{\infty} n_{BCS}(\epsilon)(-f'(\epsilon - eV)) d\epsilon, \tag{9.2}$$

where

$$G_{NN} = \frac{\alpha N_N(0) N_S(0) |\tau|^2}{e} \tag{9.3}$$

is the tunneling conductance of the contact when the superconductor is in the normal state, i.e., at $T > T_c$, and f' is the derivative of the Fermi function. Hence, the differential conductance is a thermally broadened image of the single-particle density of states of the superconductor. This simple interpretation has been employed since the pioneering experiment of Giaever [4] performed in order to test the predictions of the BCS theory and to measure the energy gap Δ (Fig. 9.3).

In the case of tunneling between a ferromagnet and a superconductor, both the density of states in the ferromagnet and the tunneling-matrix element are spin-dependent. As all quasiparticle tunneling events are independent of each other, the total current I_{FS} can be simply written as the sum of spin-up and spin-down currents, each given by (9.1) with N_N and τ replaced by their spin-dependent counterparts. This means that (9.2) remains valid, and no qualitative change is observed. The picture changes, however, when spin degeneracy in the superconductor is lifted by an externally applied magnetic field. The sharply peaked density of states in the superconductor acquires an additional Zeeman splitting of the spin-up and -down populations, and the differential conductance is now given by

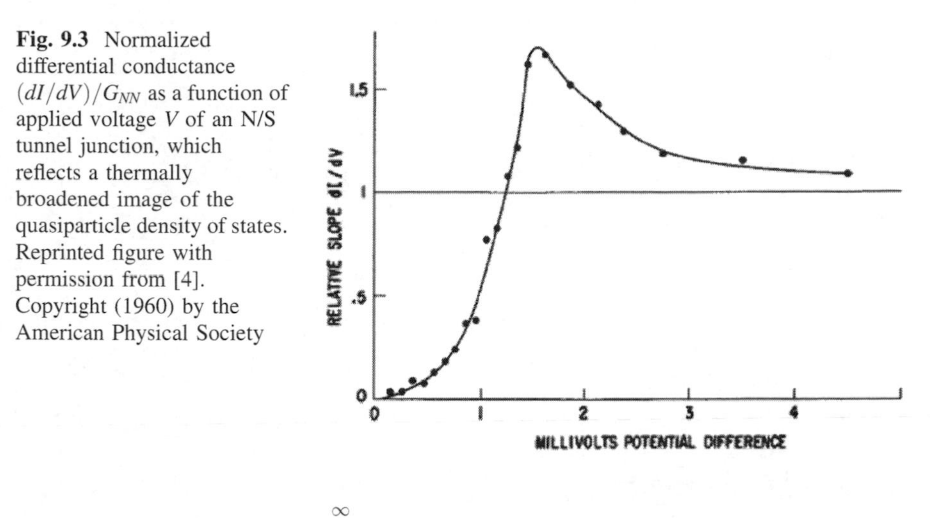

Fig. 9.3 Normalized differential conductance $(dI/dV)/G_{NN}$ as a function of applied voltage V of an N/S tunnel junction, which reflects a thermally broadened image of the quasiparticle density of states. Reprinted figure with permission from [4]. Copyright (1960) by the American Physical Society

$$\frac{dI_{\mathrm{FS}}}{dV} = G_{\mathrm{FN}}^{\uparrow} \int_{-\infty}^{\infty} n_{\mathrm{BCS}}(\epsilon)[-f'(\epsilon - eV + g\mu_{\mathrm{B}}H)]d\epsilon$$

$$+ G_{\mathrm{FN}}^{\downarrow} \int_{-\infty}^{\infty} n_{\mathrm{BCS}}(\epsilon)[-f'(\epsilon - eV - g\mu_{\mathrm{B}}H)]d\epsilon, \tag{9.4}$$

where

$$G_{\mathrm{FN}}^{\uparrow,\downarrow} = \frac{\alpha N_{\mathrm{F}}^{\uparrow,\downarrow}(0) N_{\mathrm{S}}(0) |\tau^{\uparrow,\downarrow}|^2}{e} \tag{9.5}$$

are the spin-resolved conductances of the contact in the normal state, and $g\mu_{\mathrm{B}}H$ is the Zeeman energy. To observe these effects, a magnetic field of typically several Tesla is required to achieve a sufficiently large Zeeman splitting. Here and in the following we neglect spin-orbit splitting. In bulk material of conventional low-temperature superconductors, such large fields exceed the critical field above which superconductivity is destroyed. This problem is solved by using superconducting films much thinner than the London penetration depth, and applying the magnetic field parallel to the film. In that configuration, the field penetrates the superconductor with strongly reduced screening currents, thus eliminating orbital pair-breaking. In Fig. 9.4, the results for a thin film tunnel junction between aluminum and nickel are shown [5]. At zero magnetic field (solid line), the data resemble the tunneling spectrum in Fig. 9.3, with a peak at the gap both for positive and negative voltages. At the largest field (dotted line), the peaks are clearly split into four sub-peaks, corresponding to the rigid shift of the spin sub-bands by the Zeeman energy. The peaks are asymmetric with respect to spin, i.e., peaks b and c, corresponding to spin down, are systematically higher than peaks a and d, corresponding to spin up. Thereby, a spin polarization $P_{\mathrm{tunnel}} = (G_{\mathrm{FN}}^{\uparrow} - G_{\mathrm{FN}}^{\downarrow})/(G_{\mathrm{FN}}^{\uparrow} + G_{\mathrm{FN}}^{\downarrow})$ can be determined, typically ranging from 10–50% for elementary ferromagnets. Spin dependent tunneling has been reviewed by Meservey and Tedrow [6].

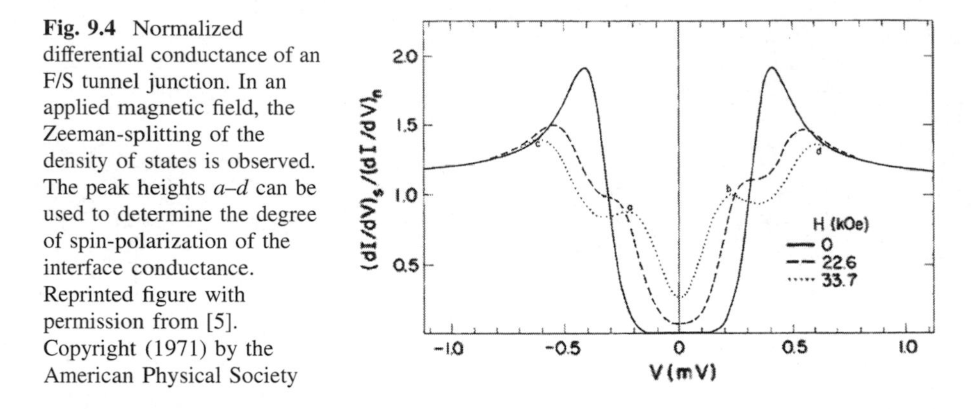

Fig. 9.4 Normalized differential conductance of an F/S tunnel junction. In an applied magnetic field, the Zeeman-splitting of the density of states is observed. The peak heights *a*–*d* can be used to determine the degree of spin-polarization of the interface conductance. Reprinted figure with permission from [5]. Copyright (1971) by the American Physical Society

9.2.2 Andreev Reflection

When an electron from a normal metal, with an energy smaller than the energy gap Δ, hits the interface to a superconductor, it cannot be transmitted as a quasiparticle. However, it can team up with a second electron to form a Cooper pair. This is described as Andreev reflection: An electron from the normal metal with an energy $\epsilon_F + \delta\epsilon$ slightly above the Fermi energy and a momentum $\mathbf{k}_F + \delta\mathbf{k}$ slightly outside the Fermi surface, is reflected as a hole of opposite spin, with energy $\epsilon_F - \delta\epsilon$ and momentum $- \mathbf{k}_F + \delta\mathbf{k}$ slightly inside the Fermi surface, on the opposite side. Thus, a net charge of $2e$ is transferred through the interface. The (almost exact) reversal of the momentum (retro-reflection) is a peculiarity of Andreev reflection, and is caused by the opposite-momentum pairing of electrons in the superconductor. It is remarkably different from the normal reflection of an electron at an interface, where only the perpendicular components of the momentum are reversed [7]. While the double charge transfer is the salient feature of Andreev reflection seen in conductance measurements, the retro-reflection aspect has been probed only in elaborate electron focussing experiments [8]. As Andreev reflection is a second-order process, requiring the transfer of two quasiparticles through the interface, it is suppressed in the presence of imperfections at the interface, like oxide layers, disorder, or Fermi wave-vector mismatch (Fig. 9.5).

Both tunneling and Andreev reflection can be treated on an equal footing by using the Bogoliubov–de Gennes (BdG) equations. Here, we will briefly describe the procedure following the classical treatment by Blonder, Tinkham and Klapwijk (BTK) [9]. The BdG equations for the electron- and hole-like quasiparticle wave functions $u(\mathbf{r})$ and $v(\mathbf{r})$ are

$$\begin{pmatrix} H_0 & \Delta(\mathbf{r}) \\ \Delta^*(\mathbf{r}) & -H_0^* \end{pmatrix} \begin{pmatrix} u(\mathbf{r}) \\ v(\mathbf{r}) \end{pmatrix} = E \begin{pmatrix} u(\mathbf{r}) \\ v(\mathbf{r}) \end{pmatrix}. \tag{9.6}$$

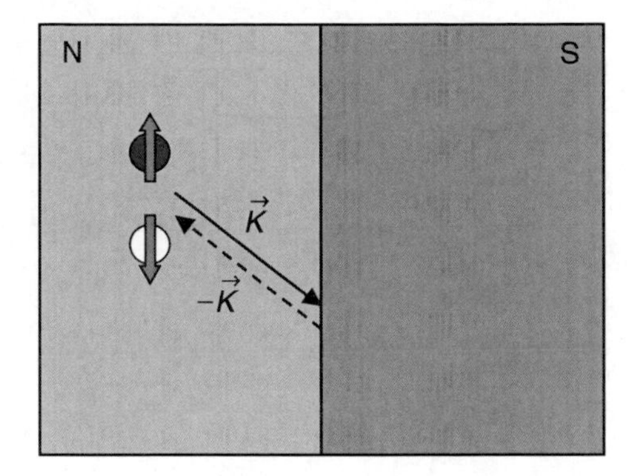

Fig. 9.5 Schematic view of Andreev reflection. An electron incident on the normal metal/superconductor interface is retro-reflected as a hole with opposite spin and momentum

Here

$$H_0 = \frac{1}{2m}\left(\frac{\hbar}{i}\nabla - \frac{eA}{c}\right)^2 + U(\mathbf{r}) - \mu \tag{9.7}$$

is the single-particle Hamiltonian in the absence of superconductivity and $\Delta(\mathbf{r})$ is the superconducting pair potential, which, in general, has to be calculated self-consistently using

$$\Delta(\mathbf{r}) = V(\mathbf{r})F(\mathbf{r}), \tag{9.8}$$

where $V(\mathbf{r})$ is the attractive pairing interaction, and the pair amplitude $F(\mathbf{r})$ is given by

$$F(\mathbf{r}) = \sum_n v_n^*(\mathbf{r})u_n(\mathbf{r})(1 - 2f_n). \tag{9.9}$$

The sum runs over all solutions of the eigenvalue problem (9.6), and f_n is the occupation probability of the eigenstate, which in thermal equilibrium is given by the Fermi function.

In the BTK treatment, a quasi-onedimensional system along the x-axis is considered, with a normal metal at $x < 0$ and a superconductor at $x > 0$. Self-consistency is ignored and Δ is set equal to its bulk value Δ_0 in the superconductor and to zero in the normal metal. This corresponds to the experimentally relevant situation of a point contact between S and N with an orifice which is much smaller than the coherence length. Furthermore, a repulsive delta-function potential of variable strength $H\delta(x)$ located at the N/S-interface is introduced, modeling the effects of finite interface transparency or interfacial disorder. The interface barrier can be expressed by the dimensionless parameter $Z = H/\hbar v_{\mathrm{F}}$, which ranges from zero for a perfectly transparent interface to infinity for a tunnel barrier.

Figure 9.6 shows a schematic view of the dispersion relations for electrons in the normal metal and quasiparticles in the superconductor near the Fermi energy. The calculation is performed by considering an incoming plane wave of unity amplitude labeled 0 in Fig. 9.6 and all the possible outgoing waves subject to energy conservation. These are (A) Andreev reflection as a hole below the Fermi

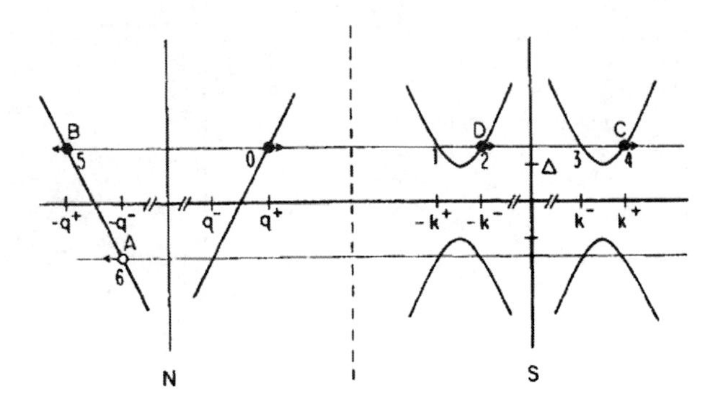

Fig. 9.6 Schematic view of the dispersion relation of a normal metal (*left*) and a superconductor (*right*). The possible transmission and reflection processes are indicated, as described in the text. Reprinted figure with permission from [9]. Copyright (1982) by the American Physical Society

surface, normal reflection as an electron (B), and transmission as a quasiparticle either without (C) or with (D) branch crossing. The amplitudes of the respective waves are found by matching the wave functions on both sides of the interface using the BdG equations. These processes are elastic and phase-coherent, which means that the reflected and transmitted waves not only have a magnitude, but also a well-defined phase relation to the incoming wave. Especially, the hole created by Andreev reflection picks up a phase factor equal to the phase of the macroscopic wave function of the superconductor. For the problem of charge transport through a single interface treated in this chapter, these phases are unimportant, but we keep them in mind as they will be crucial for the discussion of Josephson effects in Sect. 9.3. Further, Andreev reflection does not take place exactly at the interface. An electron wave incident below the superconducting gap enters as an evanescent wave, decaying roughly over the length scale ξ_S. This will be discussed further in Sect. 9.4.

After obtaining the energy-dependent transmission and reflection probabilities $A(\epsilon)$, $B(\epsilon)$, $C(\epsilon)$ and $D(\epsilon)$ from the solution of the BdG equations, we are left with the calculation of the charge current for a given voltage. As current is conserved through the interface, the calculation can be performed on either side of the interface, and is conveniently done on the normal metal side, where there is no supercurrent and only electrons contribute. The result is the integral

$$I_{\mathrm{NS}} = 2N(0)ev_{\mathrm{F}}S \int_{-\infty}^{\infty} [f(\epsilon - eV) - f(\epsilon)][1 + A(\epsilon) - B(\epsilon)]d\epsilon, \tag{9.10}$$

where $f(\epsilon)$ is the Fermi function, $N(0)$ the density of states at the Fermi energy per spin direction, v_{F} the Fermi velocity, S the cross section of the contact, and ϵ the energy relative to the chemical potential of the superconductor. The first factor of the integrand is the difference in the occupation of states on the left and right side of the interface, whereas the second factor includes a unity term for the incoming wave, plus the probability of Andreev reflection as a hole, adding to the right-going current, minus the probability of normal reflection.

In Fig. 9.7, the differential conductance (dI/dV) normalized to the normal state resistance $R_{\mathrm{N}} = (1 + Z^2)/(2N(0)e^2v_{\mathrm{F}}S)$ is shown. For high interface transparency

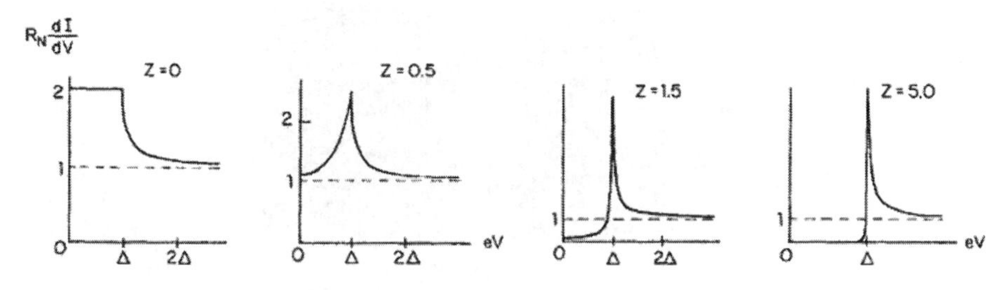

Fig. 9.7 Normalized differential conductance of a S/N-interface, for different values of the interface barrier parameter Z. Reprinted figure with permission from [9]. Copyright (1982) by the American Physical Society

($Z = 0$), a doubling of the subgap conductance is predicted, corresponding to the intuitive picture that Andreev reflection transfers a charge of $2e$ in a single event. At higher energy, where transfer of single quasiparticles dominates, the differential conductance drops to the normal state value. Upon decreasing transparency (larger Z), Andreev reflection gets supressed compared to quasiparticle transmission, as it requires the transfer of two particles through the interface. For low transparency junctions ($Z \to \infty$), the tunneling limit is recovered.

The spectroscopic investigation of Andreev reflection is a bit more complicated than tunnel junctions. For the latter, it is natural to assume that the applied voltage essentially drops at the high-resistance tunnel barrier. However, for Andreev reflection, high-transparency, i.e., low-resistance, contacts are needed. To insure a well defined voltage drop at the interface, contacts of very small cross-section between bulk materials must be formed. The ideal case is a ballistic point contact, or Sharvin contact, where the dimensions of the contact are smaller than the electron mean free path in the material [10]. The simplest way of forming such a contact is gently pressing a needle or wedge of one material onto the other, and relying on the surface roughness for the creation of a sufficiently small contact. A more elaborate way is to use nanofabricated pores in inorganic membranes, e.g., silicon nitride. After forming the small pore by lithography and subsequent reactive ion etching, the membranes are coated with two different materials from either side, thus forming the heterocontact. Both methods have been used for comparative investigations of Andreev reflection in N/S and F/S contacts [11, 12] (Fig. 9.8).

In F/S contacts, Andreev reflection is suppressed by the lack of spin-reversed partners [13]. This can be taken into account in the framework of the BTK model a posteriori by a simple phenomenological argument [12]: Andreev reflection is possible only for the fraction of incoming electrons which find a spin reversed partner. For the others, only quasiparticle tunneling is possible. Therefore, the total current can be written as $I_{\text{FS}} = P \times I_{\text{polarized}} + (1 - P) \times I_{\text{unpolarized}}$, where P is a phenomenological parameter describing the degree of spin polarization. The unpolarized current is calculated using the ordinary BTK model, while for the polarized current, the Andreev reflection coefficient $A(\epsilon)$ is set to zero, and the other coefficients are renormalized to ensure current conservation. Several authors

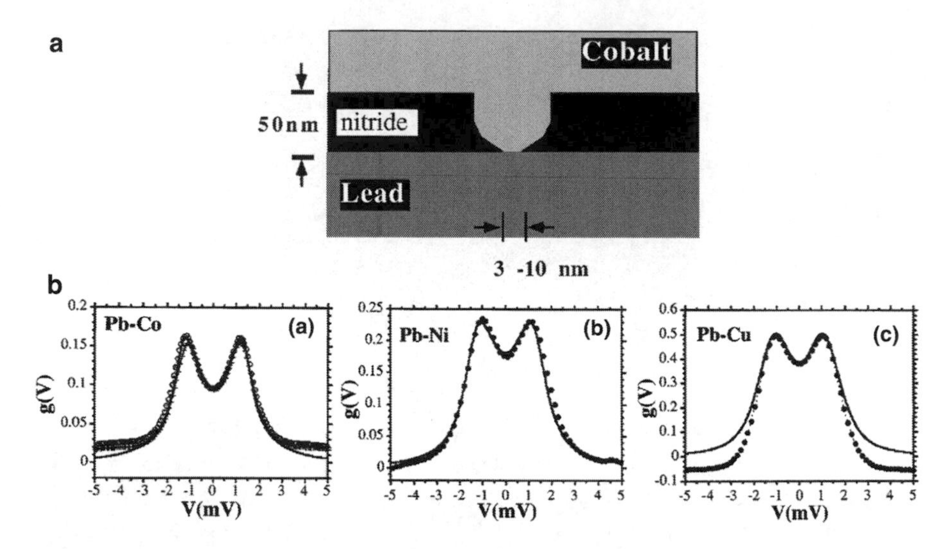

Fig. 9.8 **a** Nanofabricated point contact between a superconductor (Pb) and different normal (Cu) or ferromagnetic (Co,Ni) metals. **b** Normalized differential conductance of these point contacts. The *solid lines* are fits to a modified BTK model (see text). Reprinted figure with permission from [10]. Copyright (1998) by the American Physical Society

have modeled their experimental data in the spirit of this simple argument with good success in terms of the quality of the fits, and the consistency of the obtained parameter P. However, the situation is unsatisfactory insofar as the physical meaning of P is not obvious. A microscopic re-derivation of the current–voltage characteristics going back to the BdG equations (or up-to-date Green's function methods, see, e.g., [14]) is required. One such example has been presented in Ref. [15], where spin polarization was modeled microscopically by spin-dependent interface transmission probabilities $t^{\uparrow}$ and $t^{\downarrow}$ replacing the spin-independent parameter Z of the BTK model. The experimental data together with the fits are shown in Fig. 9.9. The data can be consistently modeled as a function of bias voltage, temperature, and magnetic field, with a single set of parameters.

9.2.3 The Proximity Effect

In the previous section, we have dealt with electronic transport through a metallic contact much smaller than the coherence length, where the equilibrium properties of the normal metal and the superconductor are not affected by their mutual presence. However, for extended interfaces, this assumption is no longer valid, and the so-called proximity effect has to be taken into account.

The microscopic origin of the proximity effect is Andreev reflection. Due to the facts that the electron and hole involved in Andreev reflection have a well-defined phase relation, and that $\delta\mathbf{k} \ll \mathbf{k}_\mathrm{F}$, the two waves will stay in phase over a long

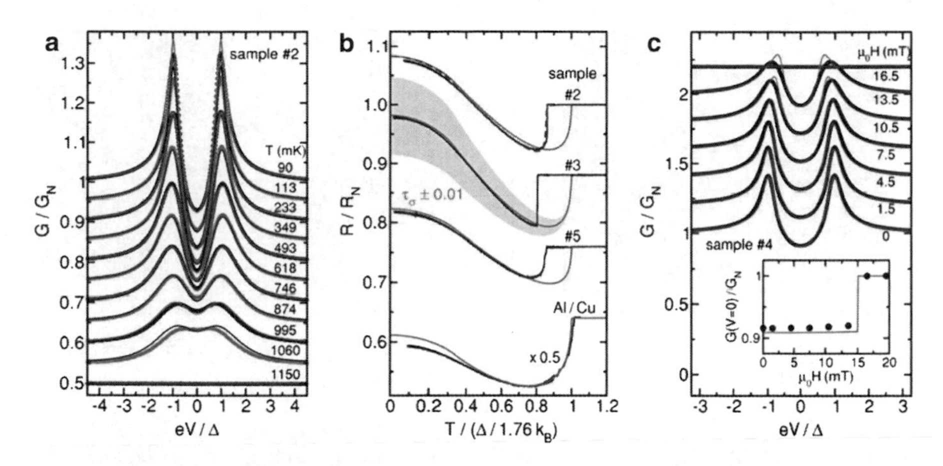

Fig. 9.9 Normalized differential conductance and fits for nanofabricated Co/Al point contacts as a function of temperature (**a**) and magnetic field (**c**). **b** Shows the normalized resistance as a function of temperature for different Co/Al contacts compared to a Cu/Al contact. Reprinted figure with permission from [15]. Copyright (2004) by the American Physical Society

distance, limited by the small $\delta\mathbf{k}$, and by inelastic scattering leading to decoherence. As all the incoming electrons below the energy gap are Andreev reflected and thereby correlated, a finite pair amplitude $F(\mathbf{r})$ is induced in the normal metal. The relevant length scale is the normal metal coherence length $\xi_N = \hbar v_F / k_B T$.

Seen from the superconducting side of the interface, Andreev reflection means that the Cooper pairs make excursions into the normal metal, thereby eventually breaking up due to the lack of the pairing interaction. In return for inducing superconductivity into the normal metal, superconductivity is therefore weakened inside the superconductor over the length scale of the coherence length $\xi_S \approx \hbar v_F / \Delta$.

A special case of the proximity effect is a thin normal metal film on top of a bulk superconductor. In addition to Andreev reflection at the N/S interface, the electrons and holes are reflected at the surface of the film. Thereby, closed trajectories including two normal reflections at the surface and two Andreev reflections at the superconductor are formed (see Fig. 9.10a). Following a simple semiclassical argument, the wave functions on the closed trajectories have to be unique, which means that the phase acquired on the loop has to be a multiple of 2π. Thus, bound states are formed, and the equilibrium density of states of the film is modified [16].

The above picture of electron–hole correlations induced by Andreev reflection has been formulated for a clean, ballistic metal, where momentum is a good quantum number. Real thin metal films, however, often have strong elastic scattering by impurities, grain boundaries, and surface roughness, leading to a diffusive motion of electrons. But even in that case, the electron and hole involved in Andreev reflection follow (nearly) time-reversed trajectories, as indicated in Fig. 9.10b. Thus, the picture of induced superconducting correlations remains

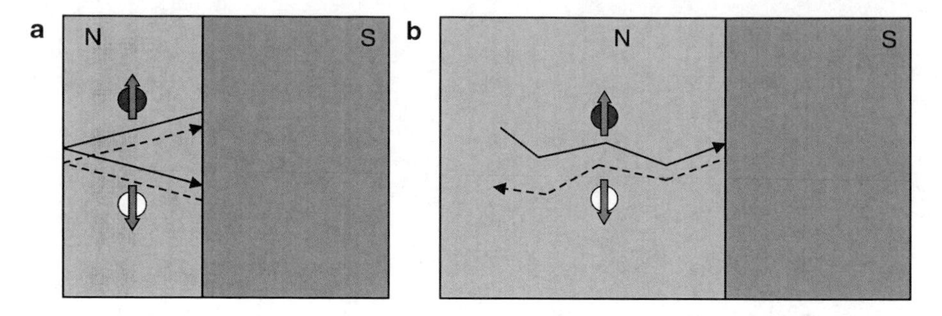

Fig. 9.10 Schematic view of the proximity effect in a ballistic thin film (**a**) and in diffusive bulk material (**b**). The *solid* and *dashed lines* represent electron and hole trajectories, respectively

valid. The relevant length scales in the diffusive regime are $\xi_N = \sqrt{\hbar D/k_B T}$ and $\xi_S = \sqrt{\hbar D/\Delta}$ in the normal metal and superconductor, respectively, where $D = v_F l/3$ is the diffusion constant, and l is the elastic mean free path.

An experimental investigation of the proximity effect in a thin normal metal film is shown in Fig. 9.11 [17]. A wedge-shaped superconducting film is coated by a normal metal layer. The local density of states is investigated at several points a–j by means of quasiparticle tunneling, using a low temperature scanning tunneling microscope. In the bilayer region a–d, the electron-hole pairs effectively diffuse back and forth between the superconductor and the surface, and the thin-film picture of bound states is valid. In this region, a superconducting gap is observed. In the proximity region e–j, the electron–hole pairs diffuse away from the superconductor, and instead of a gap, only a small depression of the density of states at low energy is seen. The transition between the two regimes is continuous.

The proximity effect is affected by ferromagnetism in a much more severe way than charge transport through the interface, as it depends on the time-reversed propagation of electron–hole pairs of opposite spin. In a ferromagnet, the Fermi

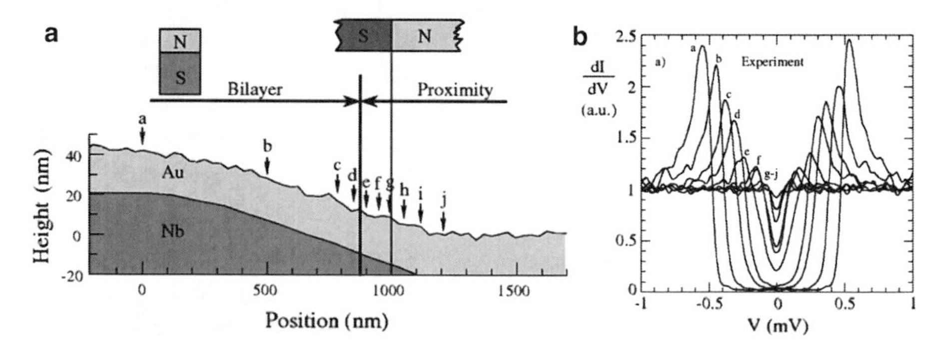

Fig. 9.11 **a** Experimental setup for the investigation of the proximity effect. A wedge-shaped superconducting niobium film is coated by a thin gold layer. The tunneling density of states is investigated at several points *a–j*. **b** Normalized tunnel conductance at points *a–j*. Figure reused with permission from [17]. Copyright (2001) EDP Sciences

wave-vector for spin up and down is split by the exchange field h_{ex}, where the splitting $\delta k_h = h_{ex}/\hbar v_F$ is a significant fraction of the Fermi wave-vector for elementary ferromagnets. Therefore, the relative phase of the electron and hole involved in Andreev reflection oscillates rapidly. As the pair amplitude $F(\mathbf{r})$ (9.9) contains pairwise products of all electron and hole wave functions, the smooth exponential decay seen in the N/S proximity effect turns into a rapid oscillatory dependence in the F/S case. Therefore, in ferromagnets the single length scale ξ_N is replaced by two different scales, the decay length ξ_{F1} of the superconducting correlations, and the oscillation period ξ_{F2}. In the ballistic limit, $\xi_{F1} = \hbar v_F/k_B T (= \xi_N)$ and $\xi_{F2} = \hbar v_F/h_{ex}$. In the diffusive limit, the momentum scattering leads to a quick averaging of all phase factors, and $\xi_{F1} = \xi_{F2} = \sqrt{\hbar D/h_{ex}}$. For diffusive films of elementary ferromagnets with exchange fields of the order of 1 eV, the proximity effect is expected to be suppressed a few nanometers away from the interface [18].

In order to study the proximity effect in ferromagnets experimentally, dilute magnetic alloys like $Pd_{1-x}Ni_x$ can be used. Palladium is a paramagnetic metal very close to a ferromagnetic instability. Upon doping with a few percent of nickel, it becomes ferromagnetic, with an exchange field that can be tuned by the nickel concentration. The density of states in a thin PdNi film on top of superconducting Nb has been studied by tunneling spectroscopy in Ref. [19], as a function of the thickness of the ferromagnetic film. The results are shown in Fig. 9.12a for two different film thicknesses. For a thin film (50 Å), a modulation of the density of states is found which is similar to the BCS-like behavior seen a normal metal film (Fig. 9.11), albeit with a much reduced modulation amplitude. However, for a thicker film (75 Å, a weaker "capsized" modulation is found, with an enhanced DOS at low energy, and two dips at the gap. For this thickness, the electrons and Andreev reflected holes acquire an average phase shift of π due to the exchange field. This turns the interference condition for the bound

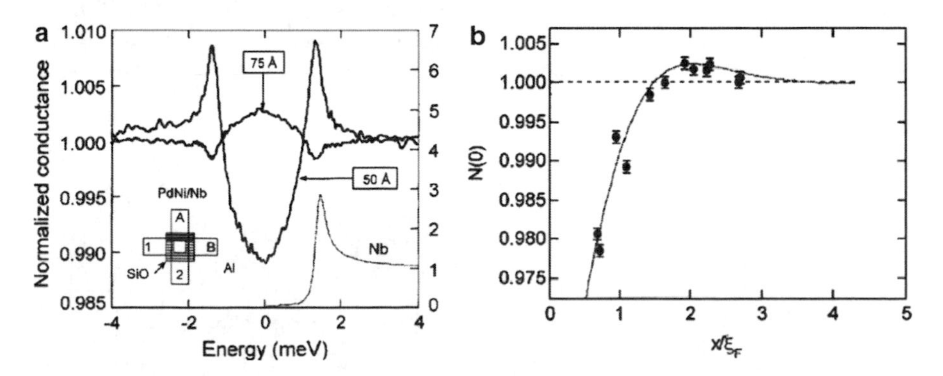

Fig. 9.12 **a** Normalized differential conductance of a dilute ferromagnetic film (*PdNi*) on top of a superconductor (*Nb*) for two different film thicknesses. **b** Density of states $N(0)$ as a function of film thickness. Reprinted figure with permission from [19]. Copyright (2001) by the American Physical Society

states from constructive to destructive (and vice-versa). The low-energy DOS as a function of film thickness is shown in Fig. 9.12b, where the oscillatory decay can be directly seen. As predicted, the decay length is the same as the oscillation period.

9.2.4 Triplet Proximity Effect?

As the proximity effect induces a weak form of superconductivity in the normal metal, it also leads to a reduction of its electric resistance, even in geometries where no net current flows through the N/S interface [20]. A number of experiments [21–24] have been designed to search for similar changes in F/S structures. In the left inset of Fig. 9.13, the setup of one of the experiments [21] is shown. A long ferromagnetic nickel wire is attached to a superconductor, and its resistance is measured as a function of temperature. At the critical temperature of the superconductor, a resistance drop ΔR is observed (main frame of Fig. 9.13). From this resistance drop, the length over which the nickel wire becomes superconducting due to proximity is estimated. It is argued that the resistance drop is too large to be consistent with the small range $\xi_{F1} = \sqrt{\hbar D/h_{ex}}$. The interpretation of this type of experiment is complicated by two factors: first, the measured resistance includes the F/S interface resistance, which is known to be modified by Andreev reflection, and second, the superconductor overlaps the ferromagnet over some length, and a current redistribution (superconducting short circuit) over that length is expected. Other experiments have been designed to overcome these problems, some showing contradictory [23] or confusing [24] results.

These experiments have stimulated an ongoing debate on whether a long-range proximity effect exists in ferromagnets, and what its nature could be. The short range of the ordinary proximity effect is a consequence of the destruction of opposite-spin pairing in conventional superconductors by the exchange field of the ferromagnet. An equal-spin (triplet) pairing might not be susceptible to the exchange field [25]. Many proposals have investigated under which circumstances

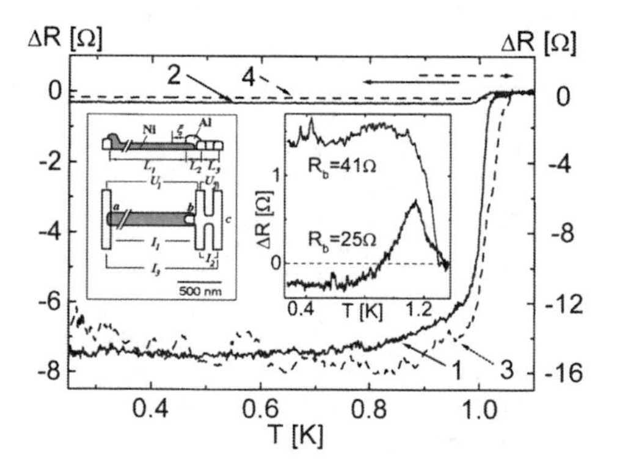

Fig. 9.13 Experimental setup (*left inset*) and results (*main frame*) of an experiment on the proximity effect in F/S structures. Reprinted figure with permission from [21]. Copyright (1999) by the American Physical Society

(inhomogeneous magnetization [26], spin-dependent scattering [27]) a triplet order-parameter component might be induced at an S/F interface, yielding a long-range proximity effect. On the experimental side, the difficulties associated with the conductance measurements described here may be overcome by looking at phase-sensitive transport, i.e., the Josephson effect, which is the subject of the next section.

9.3 Double Interfaces I: S/F/S Josephson Junctions

In the previous section, we have seen how the weak superconductivity induced by the proximity effect in a normal metal manifests itself in the single particle excitation spectrum and the conductance near an N/S or F/S interface. In this section, we will see that in an S/N/S or S/F/S junction, the same mechanism leads to the occurrence of a dissipationless supercurrent. This is the DC-Josephson effect, which was originally predicted for superconductor/insulator/superconductor junctions [28], and occurs generally in any type of weak link between two superconductors.

9.3.1 Critical Current and π-State

The microscopic mechanism of the Josephson effect in an S/N/S junction is the formation of Andreev bound states. The situation is similar to the proximity effect in a thin film: An electron from the normal metal gets reflected as a hole from the right superconductor, travels to the left, and gets reflected back as an electron. Again, a standing wave is formed, i.e., an Andreev bound state. This state carries an equilibrium supercurrent from right to left. However, for each right-going electron, there is an electron of equal energy going left, such that there is a degenerate Andreev bound state carrying a supercurrent in the opposite direction, i.e., no net supercurrent flows. Here, it becomes important that during Andreev reflection, the hole picks up a phase factor equal to the global phase of the superconducting wave function, as mentioned in Sect. 9.2.2. When there is phase difference between the superconductors on both sides of the junction, the degeneracy of the left- and right-going Andreev bound states is lifted, and a net supercurrent begins to flow. This supercurrent I_S is 2π-periodic in the phase difference $\phi_2 - \phi_1$, and in the simplest case obeys the sinusoidal Josephson relation

$$I_S = I_c \sin(\phi_2 - \phi_1), \tag{9.11}$$

where I_c is the critical current. In general, the current–phase-relation of S/N/S or S/F/S junctions depends on the spectrum of Andreev bound states, and may be strongly non-sinusoidal (see, e.g. [29]).

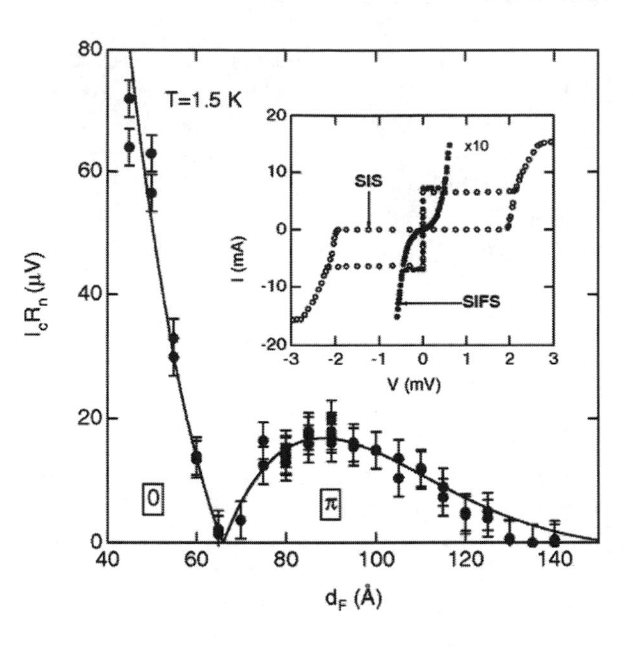

Fig. 9.14 Critical current of an S/F/S Josephson junction as a function of the thickness of the ferromagnetic layer. Reprinted figure with permission from [31]. Copyright (2002) by the American Physical Society

For S/F/S junctions, the phase differences of the single-particle wave functions between the electron and Andreev reflected hole introduced by the exchange splitting of the wave vector modify the resonance condition for the Andreev bound states in the same way as they did for the proximity effect. This means that the Josephson current in an S/F/S junction has the same oscillatory dependence on film thickness as the density of states of an F/S bilayer [30]. Especially for certain thicknesses, the direction of the Josephson current is reversed compared to the normal situation. In other words, the Josephson relation (9.11) acquires a phase shift of π.

The critical current of S/F/S junctions has been investigated by several groups [31–33], again using dilute ferromagnetic alloys. The results of such an experiment, obtained again for a dilute PdNi alloy, is shown in Fig. 9.14. The critical current as a function of ferromagnetic film thickness shows a steep decrease, drops to zero, and then recovers to a local maximum, before it disappears at large thickness. This is in good agreement with the theoretical prediction for the magnitude of the critical current (solid line), and it is very tempting to interpret the second maximum as reflecting the π-state of the junction. However, it should be noted that in this type of experiment, the applied current is controlled, and the phase difference across the Josephson junctions adjusts itself to the current, but can not be measured directly. Strictly speaking, this type of experiment can not confirm the reversed sign of the current–phase-relation predicted for the π state. When the exchange field of the ferromagnet h_{ex} is not much larger than the critical temperature T_c of the superconductor, temperature itself is a relevant energy scale, and the length scales of the proximity effect are $\xi_{F1,2} = \sqrt{\hbar D/(\sqrt{h_{\mathrm{ex}}^2 + (\pi k_{\mathrm{B}} T)^2} \pm \pi k_{\mathrm{B}} T)}$. This makes the zero-to-π transition observable not only for different samples as a function of film thickness, but also as a function of temperature for a single sample [32, 33].

The observed fast oscillatory decay of the Josephson current is fully consistent with the picture of the short-range singlet proximity effect described in the previous section. Recently, experiments on Josephson junctions incorporating CrO_2 as ferromagnet have been reported [34]. CrO_2 is a half-metallic ferromagnet, which means that only one spin species is present at the Fermi energy. In this material, the singlet proximity effect should be completely absent. However, a finite Josephson supercurrent has been observed, giving strong evidence that an induced triplet proximity effect (Sect. 9.2.4) might be present. A Josephson effect based on induced triplet superconductivity in half-metallic ferromagnets has been explicitly predicted in [27].

9.3.2 Spontaneous Currents in Loops

Phase-sensitive experiments on Josephson junctions can be performed by incorporating the junction into a superconducting loop. In this case the magnetic flux Φ through the loop controls the phase difference across the junction, where one flux quantum $\Phi_0 = h/2e$ corresponds to a phase shift of 2π. Similar to the Andreev bound states, the uniqueness of the macroscopic wave function of the superconductor requires that the overall phase shift going once around the loop is an integer multiple of 2π. If the applied magnetic flux is zero, this condition is fulfilled for ordinary Josephson junctions. However, for π-junctions, the loop has to generate an additional phase shift of π to satisfy the condition [30, 35]. This is accomplished by generating a spontaneous supercurrent, which is just large enough to produce an additional half flux quantum in the loop, as indicated in Fig. 9.15a.

In Fig. 9.15b, an experiment demonstrating the spontaneous flux of a π-junction is shown. Superconducting niobium loops incorporating either conventional or π-junctions, the latter again based on dilute PdNi alloy, were nanofabricated on top of a Hall cross etched out of a GaAs-based twodimensional electron gas. The Hall voltage V_{H} measures the total magnetic flux through the loop, including both the applied flux and the flux generated by the supercurrent in the loop. The upper panel shows the generated flux as a function of temperature for a normal and a π-junction. In the latter, upon decreasing the temperature below the critical temperature of the niobium ring, a spontaneous flux of $\Phi_0/2$ is observed, while no flux is seen in the zero-junction. When a flux of $\Phi_0/2$ is applied externally, the situation is reversed. Now the π-junction is satisfied, but the zero-junction has to generate an additional half flux quantum to make the macroscopic wave function unique.

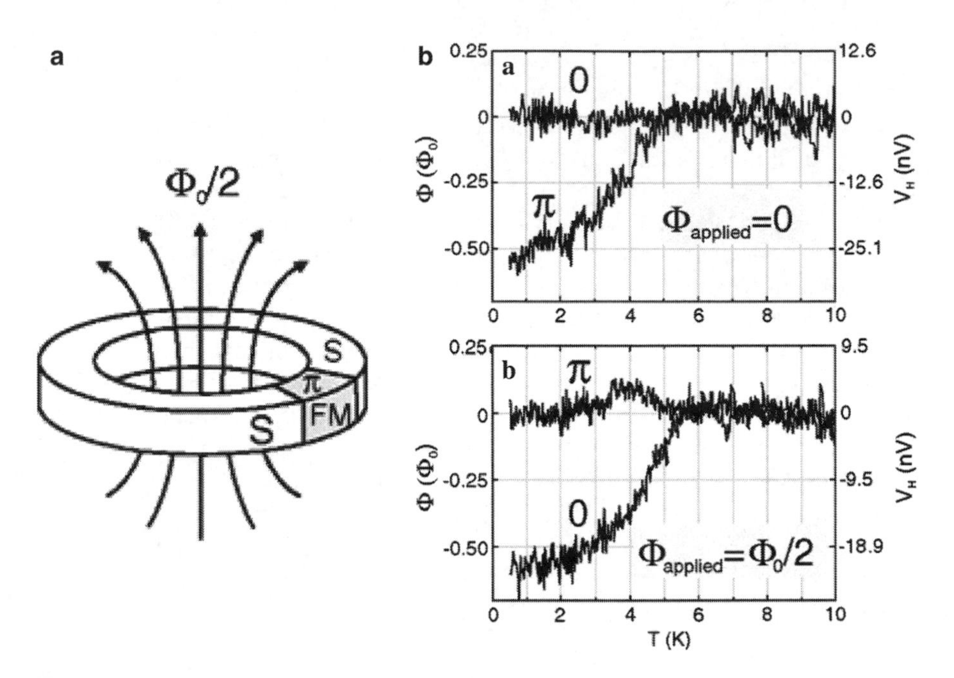

Fig. 9.15 **a** A superconducting loop incorporating a π-junction. **b** Generated flux of the π-loop, compared to an ordinary loop, as a function of temperature for different externally applied flux. Reprinted figure with permission from [36]. Copyright (2004) by the American Physical Society

9.4 Double Interfaces II: F/S/F Structures

The investigation of transport in F/S/F structures can be divided into two aspects, non-equilibrium effects due to quasi particle injection above the energy gap, and coherent non-local processes below the energy gap.

9.4.1 Non-Equilibrium Effects

Non-equilibrium effects in F/S/F structures are closely related to two independent lines of research, spin imbalance in F/N/F systems and charge imbalance in N/S/N structures. We will describe both as a precursor to understanding the main subject.

The non-equilibrium situations we are interested in occur when there are different types of charge carriers in our structures, the two spin species in the case of F/N/F structures, and the Cooper pairs and quasi-particles in the case of N/S/N systems. Each species is characterized by its own chemical potential, and in thermal equilibrium, these are equal. However, if a steady state current with an imbalance between the different species is injected, then the chemical potentials are driven out of equilibrium. The amount of non-equilibrium is governed by the injection and decay rates, and generally becomes larger with decreasing sample volume. Therefore, non-equilibrium situations are easily realized in nanostructures, and have a tendency to become a nuisance to the observation of other effects.

If a charge current I_q is passed through a normal metal, it is carried equally by spin-up and spin-down electrons. In a ferromagnet, however, the current is spin-polarized due to the different conductivities of the spin-up and spin-down sub-bands. At an F/N interface, this spin-polarized current is injected into the normal metal, leading to an accumulation of one spin species, and a depletion of the other. This spin imbalance is equivalent to a non-equilibrium magnetization, and splits the chemical potential of the two spin species. The injection rate in this case is given by the current polarization in the ferromagnet, whereas the decay is governed by spin-orbit scattering. The latter transfers the angular momentum associated with the non-equilibrium magnetization to the crystal lattice. The typical relaxation length, the spin-diffusion length λ_s, is a few 100 nm in thin films of copper or aluminum.

The detection of non-equilibrium situations can be done by means of a non-local voltage detection scheme, shown in Fig. 9.16 for the case of spin imbalance in an F/N/F structure [37]. One ferromagnetic contact is used to inject a current, creating the non-equilibrium. A second ferromagnetic contact is attached as a voltage detector outside the current path. The magnetizations of injector and detector can be switched either parallel or antiparallel, making the detector more sensitive to the chemical potential of either spin up or spin down, where up and down are defined by the injector magnetization. The voltage is measured relative to a point far to the right side of the structure, where the non-equilibrium has relaxed. This setup is referred to as a non-local spin-valve geometry.

In the case of an N/S/N structure with tunnel junctions, current is injected selectively in the form of quasiparticles, as described in Sect. 9.2.1. These quasiparticles have to recombine into Cooper pairs inside the superconductor, emitting their extra energy in the form of phonons. This process is slow at low temperatures, and in a small superconducting volume the quasiparticle distribution can differ significantly from the Fermi function assumed in (9.1). Charge imbalance has been

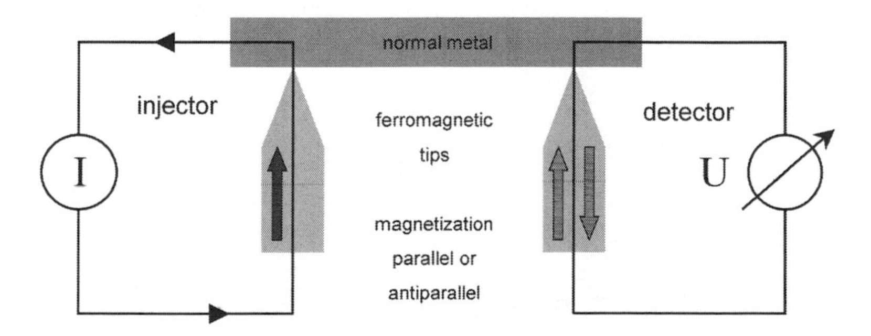

Fig. 9.16 Schematic view of a non-local spin-valve, designed to measure spin-imbalance in a normal metal. Two ferromagnetic leads are in contact with a normal metal wire. One contact is used to inject a spin-polarized current, while the second is used as a voltage detector

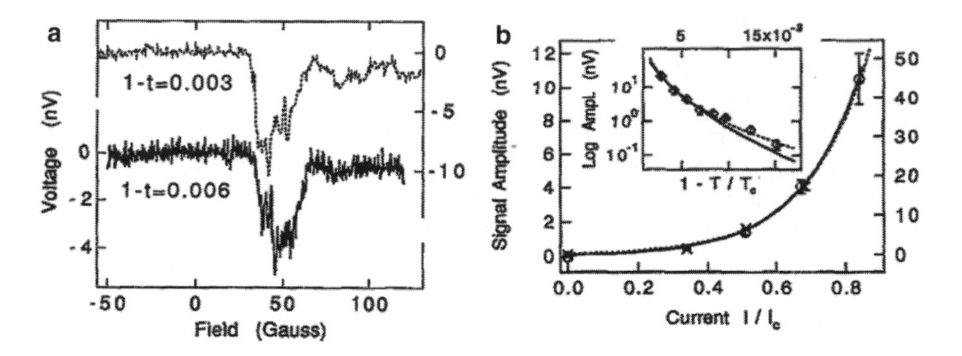

Fig. 9.17 **a** Non-local voltage as a function of the applied magnetic field in an F/S/F spin valve for two temperatures close to T_c. t is the reduced temperature T/T_c. **b** Spin-dependent signal amplitude as a function of injector current (*main frame*) and reduced temperature (*inset*). Figure reused with permission from [40] Copyright (1994) American Institute of Physics

investigated by Clarke and Tinkham [38, 39]. Here, a second tunnel junction has been used to selectively probe the chemical potential of quasiparticles relative to the chemical potential of the Cooper pairs. The charge imbalance relaxation length λ_{Q^*} is typically of the order of 10 μm.

In F/S/F structures, both spin- and charge-imbalance effects combine [40]. The non-local detector voltage for an F/S/F spin-valve structure is shown in Fig. 9.17a as a function of applied magnetic field for two different temperature close to the critical temperature T_c of the superconductor. The magnetic field is swept from negative to positive values. The voltage remains constant up to a field of about 30 G, above which it exhibits a dip. At larger field, it returns to its initial value. The magnetic-field dependence is explained by the hysteretic switching of the injector and detector magnetizations at their individual coercive fields. In the "dip" region around 50 G, the alignment is antiparallel, while in the flat regions, it is parallel. The amplitude of the dip is a measure of the imbalance of spin-up and spin-down chemical potentials. In Fig. 9.17b, the dip amplitude is shown as a function of injector current (main frame) and reduced temperature $1 - T/T_c$ (inset). The signal exhibits a non-linear dependence on injector current, and quickly decays as the temperature is lowered below the critical temperature. Both features differ from F/N/F structures, where the current dependence is linear, and the signal is (almost) independent of temperature. These observations are related to the strong temperature and energy dependence of both spectral and scattering properties of quasiparticles in the superconductor.

9.4.2 Crossed Andreev Reflection

In Sect. 9.2.2, we have already mentioned that an electron at sub-gap energies enters the superconductor as an evanescent wave over the length scale ξ_S, and that

Cooper pair formation, i.e., Andreev reflection, takes place inside the superconductor. This spatial dependence opens the possibility to observe non-local effects in a geometry where two N/S or F/S interfaces are present at a distance of the order of ξ_S. In addition to local Andreev reflection as a hole or normal reflection as an electron, the evanescent electron can be transmitted to the second interface, or even be Andreev reflected as a hole to the second contact. The latter process has been labeled "crossed Andreev reflection" (CAR) [41]. The investigation of CAR is interesting for two reasons: First, as the incoming electron and the non-locally reflected hole are "linked" by the superconducting order parameter, CAR can be used as a powerful tool to investigate unconventional superconductors [42]. The second motivation is that a Cooper pair is a quantum-mechanically entangled two-particle state. CAR separates the two particles spatially, creating a non-locally entangled Einstein–Podolsky–Rosen pair. CAR as a mechanism for a solid-state entangler might become a useful resource for quantum information processing.

Experimentally, evanescent quasiparticle propagation has been investigated both in N/S/N [43] and F/S/F [44, 45] structures. In Fig. 9.18a, the setup of the experiment in Ref. [45] is shown. The samples consist of a superconducting aluminum bar connected to several ferromagnetic iron wires. The non-local spin-valve scheme of Fig. 9.16 has been used, as for the investigation of non-equilibrium effects described in the previous section. Here, however, the contacts are highly transparent, diffusive point contacts. At low temperature and energy, current injection into the contact is mainly by Andreev reflection, i.e., the above mentioned non-equilibrium effects of quasiparticles are neglectable. While the transmission of an evanescent electron from contact A to B conserves spin, the hole created by CAR has opposite spin. Therefore, the former has a higher probability when the electrode magnetizations are parallel, whereas the latter is preferred for antiparallel alignment. This results in a signal difference ΔR_{AB} between parallel and antiparallel alignment, similar to the spin-dependent dip

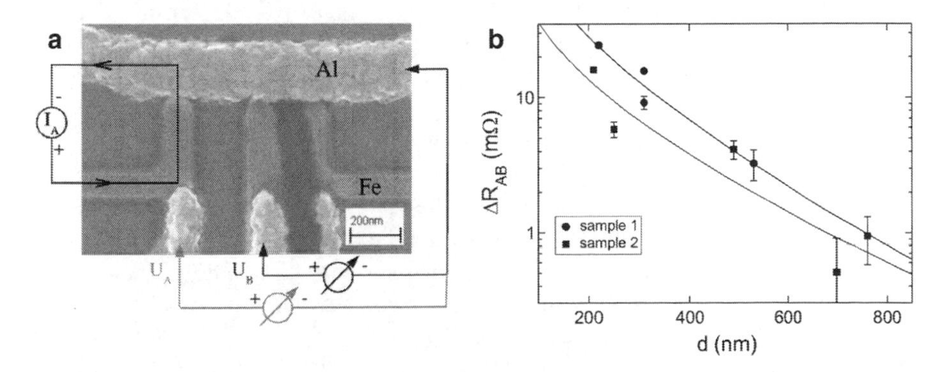

Fig. 9.18 **a** A non-local superconductor–ferromagnet spin valve used to investigate crossed Andreev reflection. **b** Spin dependent non-local resistance difference as a function of contact distance. Reprinted figure with permission from [45]. Copyright (2004) by the American Physical Society

seen in Fig. 9.17a. Figure 9.18b shows ΔR_{AB} for two different samples as a function of the distance d between the contacts. The signal decays on the length scale of $\xi_{\mathrm{S}} \approx 150$ nm, as expected for evanescent waves. This decay length is different from the typical length scales for non-equilibrium spin relaxation (≈ 500 nm), and charge imbalance relaxation ($\approx 5{,}000$ nm), obtained for the same samples at higher temperatures. The precise quantitative contribution of CAR, evanescent electron transmission, and non-equilibrium quasiparticles in F/S/F spin-valves is the subject of ongoing investigations.

9.5 Beyond Conductance: Noise and Correlations

In the previous sections, the discussion of transport properties has been restricted to (differential) conductance, i.e., the relation of the average current to the applied voltage. In general, however, due to thermal fluctuations and the discreteness of the electron charge, the instantaneous current at a given moment will deviate from the time-averaged value [46]. In other words, the current is noisy. Noise can be quantified by current–current correlation functions of the form

$$S_{\mathrm{AB}}(\tau) = \langle \Delta I_{\mathrm{A}}(t) \Delta I_{\mathrm{B}}(t+\tau) \rangle, \tag{9.12}$$

where $\Delta I_{\mathrm{A,B}}(t) = I_{\mathrm{A,B}}(t) - \langle I_{\mathrm{A,B}}(t) \rangle$ is the instantaneous deviation from the average current through a contact A or B. This definition includes both noise, i.e., the auto-correlation function S_{AA} involving only one contact, as well as cross-correlations for different contacts.

In tunnel junctions at low temperature, noise is entirely due to the uncorrelated tunneling of individual electrons, i.e., the current consists of a statistically distributed series of δ-function peaks. This shot-noise has a magnitude $S = 2eI$ in the limit of low tunnel-barrier transparency. For superconducting structures, however, Andreev reflection leads to a correlated tunneling of pairwise electrons. This is expected to result in a doubling of shot-noise, $S = 4eI$.

In Fig. 9.19a, an experimental setup for the detection of shot-noise in a contact between a superconductor and a two-dimensional electron gas based on InAs is shown [47]. In the experiment, a current is injected into the contact, and the voltage fluctuations are measured by means of a sensitive broad-band amplifier and a spectrum analyzer. From these data, the current fluctuations are calculated. The results are shown in Fig. 9.19b as a function of injector current. The noise varies linearly with injector current, and changes slope at some finite current, indicated by arrows. At this point, the voltage reaches the superconducting gap. Even though the slopes do not exactly correspond to the $2eI$ and $4eI$ mentioned above, the steeper slope at low current can be explained by the double charge transfer of Andreev reflection, whereas the smaller slope at high current is due to uncorrelated tunneling of single quasiparticles.

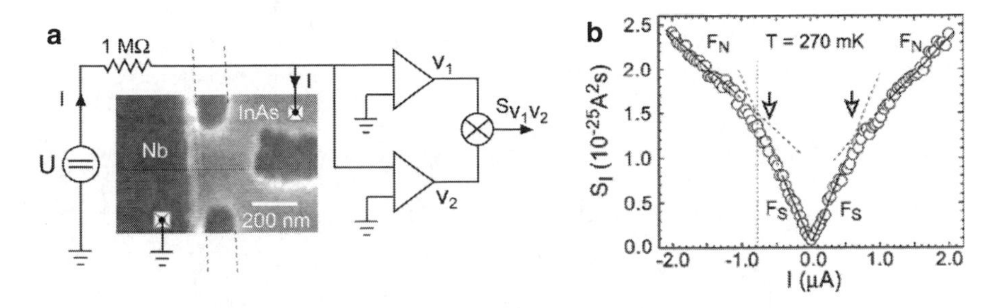

Fig. 9.19 **a** Setup for the detection of noise in a superconductor–semiconductor junction. **b** Shot noise as a function of injector bias. Reprinted figure with permission from [47]. Copyright (2005) by the American Physical Society

The structure shown in Fig. 9.19a has been designed as a two-branch beam splitter. Even though only shot-noise, i.e., current auto-correlation, in one arm was investigated, this type of structure is intended to detect cross-correlations between the two arms. Especially, positive correlations are expected when the electron and hole involved in Andreev reflection are transmitted through different arms of the beam-splitter, similar to CAR described in Sect. 4.2 in Chap. 4. No experiments have been reported on noise in F/S contacts so far, but spin-selection by ferromagnets is expected to be a valuable tool for the investigation of current correlations due to non-locally entangled pairs produced by crossed Andreev reflection.

References

1. Lévy, L.-P.: Magnetism and Superconductivity. Springer, Berlin (2000)
2. Tinkham, M.: Introduction to Superconductivity. McGraw-Hill, New York (1975)
3. Bardeen, J., Cooper, L.N., Schrieffer, J.R.: Phys. Rev. **108**, 1175 (1957)
4. Giaever, I.: Phys. Rev. Lett. **5**, 147 (1960)
5. Tedrow, P.M., Meservey, R.: Phys. Rev. Lett. **26**, 192 (1971)
6. Meservey, R., Tedrow, P.M.: Phys. Rep. **238**, 173 (1994)
7. Andreev, A.F.: Sov. Phys. JETP **19**, 1228 (1964)
8. Benistant, P.A.M., Kempen, H.v., Wyder, P.: Phys. Rev. Lett. **51**, 817 (1983)
9. Blonder, G.E., Tinkham, M., Klapwijk, T.M.: Phys. Rev. B **25**, 4515 (1982)
10. Jansen, A.G.M., van Gelder, A.P., Wyder, P.: J. Phys. C Solid Struct. Phys. **13**, 6073 (1980)
11. Upadhyay, S.K., Palanisami, A., Louie, R.N., Buhrman, R.A.: Phys. Rev. Lett. **81**, 3247 (1998)
12. Soulen, R.J.J., Byers, J.M., Osofsky, M.S., Nadgorny, B., Ambrose, T., Cheng, S.F., Broussard, P.R., Tanaka, C.T., Nowak, J., Moodera, J.S., Barry, A., Coey, J.M.D.: Science **282**, 85 (1998)
13. de Jong, M.J.M., Beenakker, C.W.J.: Phys. Rev. Lett. **74**, 1657 (1995)
14. Belzig, W., Wilhelm, F.K., Bruder, C., Schön, G., Zaikin, A.D.: Superlattices Microstruct. **25**, 1251 (1999)
15. Pérez-Willard, F., Cuevas, J.C., Sürgers, C., Pfundstein, P., Kopu, J., Eschrig, M., Löhneysen, H.v.: Phys. Rev. B **69**, 140502 (2004)
16. de Gennes, P.G., Saint-James, D.: Phys. Lett. **4**, 151 (1963)

17. Moussy, N., Courtois, H., Pannetier, B.: Europhys. Lett. **55**, 861 (2001)
18. Buzdin, A.I.: Rev. Mod. Phys. **77**, 935 (2005)
19. Kontos, T., Aprili, M., Lesueur, J., Grison, X.: Phys. Rev. Lett. **86**, 304 (2001)
20. Courtois, H., Charlat, P., Gandit, P., Mailly, D., Pannetier, B.: J. Low. Temp. Phys. **116**, 187 (1999)
21. Petrashov, V.T., Sosnin, I.A., Cox, I., Parsons, A., Troadec, C.: Phys. Rev. Lett. **83**, 3281 (1999)
22. Giroud, M., Courtois, H., Hasselbach, K., Mailly, D., Pannetier, B.: Phys. Rev. B **58**, R11872 (1998)
23. Aumentado, J., Chandrasekhar, V.: Phys. Rev. B **64**, 054505 (2001)
24. Giroud, M., Hasselbach, K., Courtois, H., Mailly, D., Pannetier, B.: Eur. Phys. J. B **31**, 103 (2003)
25. Bergeret, F.S., Volkov, A.F., Efetov, K.B.: Rev. Mod. Phys. **77**, 1321 (2005)
26. Bergeret, F.S., Volkov, A.F., Efetov, K.B.: Phys. Rev. B **86**, 4096 (2001)
27. Eschrig, M., Kopu, J., Cuevas, J.C., Schön, G.: Phys. Rev. Lett. **90**, 137003 (2003)
28. Josephson, B.D.: Phys. Lett. **1**, 251 (1962)
29. Golubov, A.A., Kupriyanov, M.Y., Il'ichev, E.: Rev. Mod. Phys. **76**, 411 (2004)
30. Buzdin, A.I., Bulaevskiı, L.N., Panyukov, S.V.: JETP Lett. **35**, 178 (1982)
31. Kontos, T., Aprili, M., Lesueur, J., Genet, F., Stephanidis, B., Boursier, R.: Phys. Rev. Lett. **89**, 137007 (2002)
32. Ryazanov, V.V., Oboznov, V.A., Rusanov, A.Y., Veretennikov, A.V., Golubov, A.A., Aarts, J.: Phys. Rev. Lett. **86**, 2427 (2001)
33. Oboznov, V.A., Bol'ginov, V., Feofanov, A.K., Ryazanov, V.V., Buzdin, A.I.: Phys. Rev. Lett. **96**, 197003 (2006)
34. Keizer, R.S., Goennenwein, S.T.B., Klapwijk, T.M., Miao, G., Xiao, G., Gupta, A.: Nature **439**, 825 (2006)
35. Bulaevskii, L.N., Kuzii, V.V., Sobyanin, A.A.: JETP Lett. **25**, 290 (1977)
36. Bauer, A., Bentner, J., Aprili, M., Della Rocca, M.L., Reinwald, M., Wegscheider, W., Strunk, C.: Phys. Rev. Lett. **92**, 217001 (2004)
37. Johnson, M., Silsbee, R.H.: Phys. Rev. Lett. **55**, 1790 (1985)
38. Clarke, J.: Phys. Rev. Lett. **28**, 1363 (1972)
39. Tinkham, M., Clarke, J.: Phys. Rev. Lett. **28**, 1366 (1972)
40. Johnson, M.: Appl. Phys. Lett. **65**, 1460 (1994)
41. Deutscher, G., Feinberg, D.: Appl. Phys. Lett. **76**, 487 (2000)
42. Byers, J.M., Flatté, M.E.: Phys. Rev. Lett. **74**, 306 (1995)
43. Russo, S., Kroug, M., Klapwijk, T.M., Morpurgo, A.F.: Phys. Rev. Lett. **95**, 027002 (2005)
44. Gu, J.Y., Caballero, J.A., Slater, R.D., Loloee, R., Pratt, W.P. Jr.: Phys. Rev. B **66**, 140507 (2002)
45. Beckmann, D., Weber, H.B., Löhneysen, H.v.: Phys. Rev. Lett. **93**, 197003 (2004)
46. Blanter, Y.M., Büttiker, M.: Phys. Rep. **336**, 2 (2000)
47. Choi, B.-R., Hansen, A.E., Kontos, T., Hoffmann, C., Oberholzer, S., Belzig, W., Schönenberger, C., Akazaki, T., Takayanagi, H.: Phys. Rev. B **72**, 024501 (2005)